W9-BVH-533

DATE DUE

HURRICANES
of the Gulf of Mexico

The Gulf of Mexico Basin and surrounding environment

HURRICANES
of the Gulf of Mexico

Barry D. Keim and Robert A. Muller

LOUISIANA STATE UNIVERSITY PRESS

BATON ROUGE

Published by Louisiana State University Press
Copyright © 2009 by Louisiana State University Press
All rights reserved
Manufactured in the United States of America
First printing

Designer: Laura Roubique Gleason
Typefaces: Whitman and Helvetica Neue
Printer and binder: Thomson-Shore, Inc.

Library of Congress Cataloging-in-Publication Data

Keim, Barry D., 1963–
    Hurricanes of the Gulf of Mexico / Barry D. Keim and Robert A. Muller.
        p. cm.
    Includes bibliographical references and index.
    ISBN 978-0-8071-3492-4 (cloth : alk. paper)   1. Hurricanes—Mexico, Gulf of—History—
20th century.   I. Muller, Robert A.   II. Title.
    QC945.K45 2009
    551.55'20916364—dc22

                                                                                2009008298

To residents along the Gulf of Mexico coast,
who persevere in the face of hurricanes
for the bounty the Gulf brings.

To my wife, Ellen, and our children, Anderson and Nathan
—BDK

To my wife, Sonni, and my children Lisa and John
—RAM

Contents

Preface

At very early ages, we both developed an unusual interest in weather. In Chalmette, Louisiana, a downstream suburb of New Orleans, Barry watched flooding rainstorms and the approaches of tropical storms and hurricanes from the Gulf. In suburban New Jersey, Bob was excited about every snowfall, making careful measurements with a ruler and fretting during northeasters over whether the snow would change to sleet, freezing rain, or rain.

From his parents, Barry heard eyewitness stories about Hurricane Betsy in 1965 with its destructive wind damage, failed levees, and widespread flooding. Betsy made landfall on Barry's second birthday, and the Keim household in Chalmette took in four feet of water. Like most Chalmette residents, they rebuilt right there. However, a few years after Betsy, Barry's father had the wisdom to raise the elevation of the new family home by adding a third story and dormer windows, in case of a major flood and the need to get to the roof for rescue. While hunkered down in the new and improved family home, Barry watched the forecasts and near misses of major hurricanes such as Camille in 1969 and Carmen in 1974. Those storms and others clearly made an impression on him, piquing his curiosity about the power of weather.

Bob experienced the western fringes of the Great New England Hurricane of 1938, watching with awe as a tall sycamore tree came down. In 1955, he drove with friends near Sandy Hook ahead of Hurricanes Connie and Diane to watch the storm surf from a high bluff overlooking the beach. Bob moved south to join the Louisiana State University faculty in 1969, and during family vacations at Pensacola Beach he had to evacuate ahead of Hurricanes Frederic in 1979, Elena in 1985, and Erin in 1995. In Baton Rouge, he experienced a weakening Hurricane Edith in 1971, when male students at LSU went outside to enjoy the brisk winds without heavy rain while their female classmates were prevented from leaving

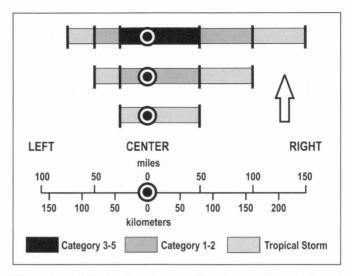

Dimensions of the tropical storm and hurricane strike model used throughout the book.

their dorms. In 1992 Hurricane Andrew left the Muller home largely undamaged but without electrical power for eight sultry days in late August and early September.

Barry and Bob first became acquainted in 1985 when Barry enrolled in Bob's introductory meteorology class. This interaction led to Barry's master's thesis on the climatology of heavy rain events in New Orleans and ultimately to his doctoral dissertation on heavy rain events across the South. After a period of teaching and serving as state climatologist in New Hampshire, Barry returned to his beloved Louisiana with a joint position at LSU as regional climatologist in the Southern Regional Climate Center and assistant professor in the Department of Geography and Anthropology. In 2003 he became Louisiana State Climatologist, charged with archiving climatic data, making the data available to the public, educating the people of the state and region on matters related to weather and climate, monitoring the impact of climate on the regional economy, and serving as a contact for climatic research. During hurricanes, he forms part of a team from LSU that advises the Governor's Office for Homeland Security and Emergency Preparedness as it attempts to mitigate damage and loss of life in Louisiana.

After the destructive strikes of Hurricanes Erin and Opal at Pensacola

Beach in 1995, Greg Stone of the Coastal Studies Institute at LSU asked Bob to research the history of tropical storm and hurricane strikes at Pensacola Beach. To evaluate strike intensities at particular beach locations, Bob developed a storm strike model with average radii of Saffir-Simpson hurricane category conditions to the left and right of the storm track, using the track and category archive data of the National Hurricane Center. The original research going back to 1901 for Pensacola Beach made him curious about the strike history at other resorts and towns along the northern Gulf Coast, and soon after Bob completed a comparative analysis from Apalachicola on the east to Galveston on the west.

Bob and Greg then extended the study to cover the subtropical coasts of the United States from Cape Hatteras to South Padre Island. The geographical and temporal patterns of strikes were so different and provocative that Bob turned to Barry for investigation of the results in relation to established calendars of ocean sea-surface temperature and atmospheric pressure patterns for the North Atlantic and Pacific Oceans. Together, Bob, Barry, and Greg completed the compilation and analysis for the Atlantic Coast from Maine to Texas.

Finally, we—Bob and Barry—focused our attention on tropical storm and hurricane strikes around the shores of the Gulf of Mexico, which became the basis for this book. Our work was further stimulated by the recurring coastal destruction and loss of life during the 2004 and 2005 seasons by major Hurricanes Charley and Wilma in southwestern Florida and by Hurricanes Ivan, Dennis, Katrina, and Rita along the northern Gulf Coast from Panama City nearly to Galveston. Barry was personally impacted when his family home in Chalmette was inundated by nine feet of Katrina floodwaters.

In this book, we highlight tropical storms and hurricanes of the Gulf of Mexico over the past 100–150 years, the period in which records are available from the National Hurricane Center. In most instances, we cover the years 1901–2008, because records from the nineteenth century are incomplete for storms that did not make landfall. For storm strikes in New Orleans, however, we extend coverage back to 1851, as events were well documented for that city. We also strive to demonstrate the unique features of the Gulf that influence the development of hurricanes, such as the loop current and its eddies, and to identify areas of the coastline that are more or less vulnerable because of the physical environment, socioeconomic environment, or both.

We begin with a synopsis of the devastating Galveston Hurricane in September 1900, which saw the greatest loss of life from a natural disaster in the history of the country. Given the rudimentary scientific understanding of hurricane development and tracks more than one hundred years ago, the very limited weather observations by ships at sea, and the inability to transmit observations from ship to shore, it is not too surprising that the Weather Bureau headquarters in Washington, D.C., expected the hurricane to be off the eastern coast of the United States rather than approaching Galveston, where on the very morning of the hurricane some residents went to the beach to watch the waves and surf.

For comparative purposes, we then review the day-by-day tracking forecasts and storm strength of Hurricane Katrina. The Katrina odyssey began over the Bahamas before the storm crossed southern Florida, emerged over the Gulf becoming a category five storm, and made landfall in Louisiana south of New Orleans as a category three, maintaining this strength during a second landfall on the Louisiana-Mississippi border at the Pearl River. More than one hundred years after Galveston, and despite the development and routine use of radar and satellite observations and advanced computer models, it was not until about forty-eight hours before landfall that very specific watches and warnings could be issued to the hardest hit areas in Louisiana and Mississippi.

Our focus then turns to the great variability of storm frequencies through the decades as we present a climatology of tropical storms and hurricanes over the Gulf relative to storms over the entire North Atlantic Basin beginning with 1901. The first category five hurricane over the entire Atlantic occurred during the 1928 season, and over the Gulf, during the 1935 season. The latter storm, which became known as the Labor Day Hurricane, swept over the Florida Keys from Atlantic waters to the Gulf, drowning several hundred construction workers at work on the new overseas highway to Key West and leaving behind a narrow path of destruction that included the demise of the overseas railroad to Key West. A unique graph shows the geographical and temporal patterns of all tropical storm and hurricane strikes at thirty locations around the Gulf from Key West to the Yucatan from 1901 through 2007.

Physical geography dictates that the development of these uncommon storms—called hurricanes in North America, typhoons over the western Pacific, and cyclones over the Indian Ocean—is restricted to rather specific regions of the tropical oceans with similar meteorological atmospheric structures. We explain how tropical storms and hurricanes develop from poorly defined tropical "waves" over warm seas and examine their typical tracks and life cycles. Along the shores of the Gulf, the geographical configurations of land and water result in very revealing storm strike patterns.

We focus on a selection of the Gulf Coast's more memorable hurricanes, such as the Cheniere Caminada Hurricane in southeastern Louisiana in 1893, the Labor Day Hurricane in the Florida Keys in 1935, and Hurricane Camille along the Mississippi Coast in 1969. Included is an overview of the tragic, stronger-and-earlier-than-expected landfall of Hurricane Audrey in 1957 in Cameron Parish, Louisiana—in terms of time and geography, just about halfway between the Galveston Hurricane and Katrina. In addition, the hurricane histories of ten cities around the Gulf from Key West, Florida, to Progreso in the Yucatan are presented as case studies over time. We address the environmental and socioeconomic impacts of these hurricanes, and we conclude by considering the prospects of hurricane events over the next several decades, in light of varying perspectives on global warming and the multidecadal cycles of oceanic warming and cooling.

We finished writing our book in the spring of 2008. While performing the final edits, Hurricane Dolly made landfall in south Texas in late July of that year. Gustav and then Ike ravaged the north-central Gulf Coast in September 2008. As a result, we added an epilogue summarizing the impacts of these storms. Coincidentally, we brought the book full circle from where we began it, on Galveston Island.

Acknowledgments

We are grateful to our families, friends, and colleagues who encouraged us to develop and complete this study. We particularly want to recognize the cartographic contributions of Clifford P. "Dupe" Duplechin and Mary Lee Eggart of the Department of Geography and Anthropology at Louisiana State University. Without their tireless efforts, preparation of many of this book's graphics would have been impossible. We also thank Gregory W. Stone, Director of LSU's Coastal Studies Institute, for encouraging our initial work in the evaluation of hurricane strikes at Pensacola Beach, Florida. We thank the professional staff at the National Hurricane Center for the open public record of advisories, forecasts, and reanalysis of all Atlantic Basin storms dating back to 1851.

At LSU Press, we wish to thank Margaret Hart Lovecraft, who first recognized the merits of this study; Laura Gleason, who managed the reproduction of our graphics; Catherine L. Kadair, who guided and exhorted us through the final stages of production; and MaryKatherine Callaway, Director of LSU Press, who made the ultimate decision to proceed. We also thank the outside reviewers, who made suggestions for improvements.

This research was supported in part by NOAA Grant No. NA08OAR4320886.

HURRICANES
of the Gulf of Mexico

| 1 |

The Galveston Hurricane of 1900

On the morning of Saturday, September 8, 1900, some residents of the seaside city of Galveston, Texas, flocked to the beach to watch the very large waves and surf that were crashing on the beaches. For late summer there was an unusually brisk wind from the north. The Weather Bureau had warned residents of an oncoming coastal storm, but there were no official Weather Bureau reports of a hurricane over the Gulf. Many of these residents had experienced mostly nondestructive coastal storms in recent decades, and apparently expectations were for just another storm that would break the late-summer doldrums and provide a bit of excitement and entertainment at the beaches. Just a few hours later, however, Galveston was devastated by a major hurricane with a record storm surge that flooded and destroyed much of the city, resulting in the greatest loss of life during any American natural disaster to the present day.

At the turn of the century, Galveston was one of the most important ports of entry in the entire country, rivaling New Orleans on the Gulf Coast. Further growth seemed assured because Texas and the Great Plains to the north were being inundated with new settlers, and commerce and the economy were growing rapidly. However, the city of Galveston was located on a barrier island, thirty miles long, but only two miles wide at most, and separated from the mainland by Galveston Bay, which is three miles wide at its narrowest point (see fig. 1.1). Three railroad viaducts and a single roadway—at the time, a wagon bridge—over the bay connected the city with the mainland, but these transportation arteries were especially vulnerable to closure during stormy weather (see fig. 1.2). Furthermore, the developing city of Houston, located more than forty miles inland, but with plans for an oceangoing ship canal connecting Houston with the Gulf and new port facilities, was becoming serious competition for Galveston, offering shippers the potential for much more efficiency and safety than at the port of Galveston.

Fig. 1.1. The Galveston area as of 1900 before the hurricane.

Fig. 1.2. This 2.1-mile wooden-plank wagon bridge was said to be the longest roadway bridge in the world when completed in 1893. It connected Galveston with the mainland but, along with three railroad trestles, was destroyed during the hurricane of 1900, making access to Galveston very difficult after the storm.

The highest point on Galveston Island was less than nine feet above the sea. There were bathhouses and pavilions at the beach, with several extending outward on pilings over the surf for more than a hundred feet. Commercial and shipping facilities were located on the mainland side of the island facing Galveston Bay. Larger homes and mansions were mostly on higher ground midway across the island along Broadway. Seeing as the city had easily survived several tropical storm and hurricane strikes in the nineteenth century, the local chief of the Weather Bureau at that time, Isaac Cline, believed that the broad, shallow slope of the Gulf bottom offshore and the extensive marshlands on the inland margins of Galveston Bay helped to make the city less vulnerable to really destructive storm surges. Cline's belief was contrary to the hurricane surge events in 1875 and 1886 at Indianola, a coastal town 115 miles southwest of Galveston with six thousand residents in 1875. The 1875 storm surge at Indianola destroyed three-quarters of the buildings, and the even higher surge in 1886 destroyed all buildings or rendered them uninhabitable. Indianola was never rebuilt.

In order to follow the development and progress of the storm that overwhelmed Galveston on Saturday, September 8, 1900, we need to understand the organization of the United States Weather Bureau at that time and the technology available for monitoring storms at sea and forecasting their tracks and intensities. The Weather Bureau had originally been administered by the Signal Corps of the United States Army, with a hierarchy similar to the military and absolute authority for all operations and forecasts centralized at headquarters in Washington, D.C. All forecasts of extreme weather events, whether tornadoes, hurricanes, or floods, had to emanate from headquarters and were then transmitted by telegraph lines to local offices such as Galveston's. Weather Bureau headquarters was reluctant to mention or forecast events that could negatively affect business activity, and the use of the term *hurricane* was restricted to rare, extremely severe tropical storm systems. In 1891 the Weather Bureau was transferred to the Department of Agriculture, since farmers and ranchers were very interested in obtaining weather reports and forecasts focusing on crop production and environmental conditions. But despite the administrative relocation, the Weather Bureau continued to preserve its military command structure.

In 1900, Cuba was still under the administration of the War Department of the United States, after the conclusion of the Spanish-American

War of 1898 and the removal of the Spanish colonial government. The Weather Bureau was not impressed by the Cuban meteorologists at Havana, who had many decades of experience with local forecasts of hurricanes over Cuba and the West Indies and also a strong following among the Cuban population. The Weather Bureau had established a new American Weather Bureau office in Havana, but the Weather Bureau chief in Washington soon became irritated by forecasts published daily by the Cuban meteorologists that conflicted with forecasts issued by headquarters in Washington. In August 1900, in the middle of the hurricane season, the Weather Bureau was successful in preventing cable transmission of all weather information regarding storm warnings by the Cuban meteorologists. Hence, reporting of tropical storms or hurricanes back to headquarters was limited to the very conservative perspectives of tropical storm development by Weather Bureau personnel, and the Cuban expertise was largely ignored.

By Friday, August 31, based on island observations entered into the telegraphic network, both American and Cuban meteorologists recognized that a weak tropical storm was moving westward over the Atlantic across the Leeward Islands east of Puerto Rico. The storm remained rather weak and disorganized as it passed over Puerto Rico, the Dominican Republic, and Haiti but arrived over southeastern Cuba somewhat

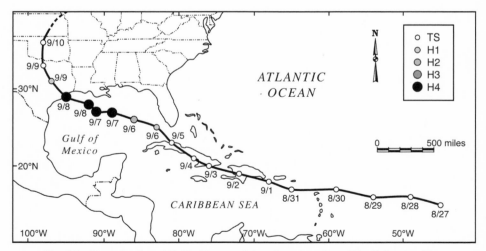

Fig. 1.3. Storm track and Saffir-Simpson intensity categories of the Galveston Hurricane of 1900 as interpreted by the National Hurricane Center.

strengthened on Monday, September 3, when twenty-four inches of rain were recorded at Santiago, Cuba (see fig. 1.3). On Wednesday, September 5, only three days before landfall at Galveston, the tropical storm swept over western Cuba moving from southeast to northwest without serious damage. Cuban meteorologists suspected hurricane status once the storm left Cuba, but Weather Bureau personnel in Cuba continued to report to Washington that the tropical storm was weakly organized.

The storm strike in Cuba was reported by telegraph to the Weather Bureau in Washington. Forecasters at headquarters anticipated that the storm would follow a traditional track and curve slowly toward the north and then northeast, cross northern Florida, emerge over the Atlantic Ocean, and then track northeastward along the Carolina and Mid-Atlantic Coasts. But real-time weather data around the Gulf and data transmission by teletype to headquarters were limited to the coastal Weather Bureau offices, and it was not possible at that time for real-time transmission of observations from ships at sea on the Gulf. Hence, after the hurricane left Cuba and moved over the southeastern Gulf, there were no observations to pinpoint the precise position of the storm, unlike Hurricane Katrina more than one hundred years later.

On Thursday, September 6, less than three days before landfall in Galveston, the Weather Bureau in Washington continued to alert shipping interests along the East Coast from the Carolinas northward of the possibility of a major storm event. However, there had been no evidence of a storm tracking across the Florida Peninsula from the Gulf to the Atlantic, and observations of brisk northerly winds at coastal Weather Bureau offices along the central Gulf should have suggested that the storm—or cyclone, a term often used for storms of tropical origin—was moving in a northwesterly direction across the central Gulf of Mexico. The Weather Bureau forecasting services in Washington did not order storm predictions or warnings for the Gulf Coast, and local Weather Bureau personnel along the Gulf Coast were not allowed to issue storm or hurricane warnings on their own.

Early on Friday morning, September 7, only about thirty-six hours before landfall, local meteorologist Cline, living only three blocks from the beach, became aware of rising water levels, storm swells from the southeast, and a thunderous surf. Cline recognized that there had to be a significant storm somewhere over the Gulf (see fig. 1.4). The Weather Bureau in Washington finally ordered "storm warnings" along the northern coasts of the Gulf, including Galveston Island, but with no mention of a hurricane threat. The weather on the island remained fair, and residents were mostly unaware of the storm approaching from the southeast. In fact, adults and children flocked to the beach to enjoy the brisk cooler winds from the north and the crashing surf.

Weather Bureau transmissions from Washington had continued to suggest that a storm was moving up the Atlantic Coast, but at 11:30 a.m. another transmission from Washington arrived in Galveston reporting that a "moderate" tropical storm was over the Gulf, south of Louisiana, and moving toward the northwest, threatening the southeastern Texas coast Friday night and Saturday. Cline's office issued a storm warning, but no hurricane warning, since issue of these warnings was still strictly limited to headquarters in Washington. It should be noted that modern analyses of the storm track and intensities by the National Hurricane Center (see fig. 1.3) indicate that the storm had exploded to category four status with the center only about 150 miles southeast of Galveston by Friday morning, with the outer bands of hurricane-force winds much closer. Later in the afternoon with fair but windy weather, Cline observed that the sea and surf were still rising and beginning to flood the ends of streets

Fig. 1.4. Dr. Isaac Cline, chief of the Weather Bureau office in Galveston during the 1900 hurricane. This photo was taken much later, during his controversial tenure in charge of the Weather Bureau Center for the Gulf Coast in New Orleans (1901–1935).

at the beach; he sent a concerned telegram to headquarters in Washington that he had never before observed such high water together with brisk offshore winds from the north.

At dawn, Saturday, September 8, the surf was even higher. Rising water from Galveston Bay, driven by the brisk northerly winds, was beginning to flood streets and wharves along the commercial waterfront there. But most city officials and the public, including Cline and the other Weather Bureau personnel, anticipated a day of stormy weather but nothing that had not been experienced before. They had no clue that they were about to experience a major hurricane strike that very afternoon. For much of the early morning, families walked or rode streetcars to the beach to watch the spectacular fountains of surf against the bathhouses and piers.

By noon, however, driving rain, rising Gulf water levels, and mountainous waves began crashing into the bathhouses and stores at the beach. The streetcars stopped operating, and the crowds at the beach were now desperately trying to make their way through the rising water and increasing winds toward Broadway and the substantial buildings along the highest ground on the island. It soon became difficult to wade through the seawater, which was laden with debris and flowing rapidly more or less from east to west, nearly parallel with the beaches. At the same time, screaming winds from the north were beginning to tear apart roofing, with shingles and slate becoming death-dealing missiles. The island was being flooded both from the Gulf, due to the rising storm surge and high waves, and from Galveston Bay, because of gale- and hurricane-force winds from the north.

The last train to arrive in Galveston on Saturday had left Houston early in the morning filled with passengers. It began the treacherous three-mile crossing on a trestle over Galveston Bay more or less on schedule. The train shuttered in the driving wind and rain with bay water almost up to the rails. The crossing to the island was successful, but the train had to stop two miles from Galveston because of a washout ahead with flooded wetlands on both sides of the tracks. More than an hour later a rescue train arrived on an adjacent track after crossing another trestle across the bay. Passengers and crew were transferred to the second train, which arrived in Galveston at 1:15 p.m. after creeping slowly through rising waters several feet deep in places. By this time water from the bay was waist deep around the train station and the Tremont Hotel in the business district.

The evidence suggests that most Galveston residents recognized that a tropical storm was approaching the city on Saturday morning. However, they had no forewarning about the potential severity of the storm, based on local experiences with weaker tropical storms in recent decades and the apparent lack of dire forecasts and warnings from Weather Bureau headquarters. Indeed, we have seen that as late as midmorning some residents were still flocking to the beach to see the waves, but by early afternoon, only two to three hours later, they were trying to flee the beach areas as streets filled with seawater, massive amounts of floating debris, and even bodies, especially those of children, carried along by the rushing current.

Firsthand accounts by survivors tell a story of hurricane winds veering from north to northeast to east and eventually the south during Saturday afternoon and evening, with a record storm surge of about twenty feet in early evening after winds shifted to the east and then the south as the storm center crossed the coast just to the west of Galveston. By late afternoon the storm surge began pushing a more than ten-foot-high and miles-long wall of debris across the island from southeast to northwest, with the wall working like a giant bulldozer bringing down still-standing houses and other structures as it shifted in from the beach (see fig. 1.5). Eyewitness stories describe the very moments when the houses began to fall into the raging floodwaters ahead of the debris wall and how survivors managed to support themselves on floating debris for hours.

When survivors began to regroup after dawn on Sunday morning, the winds were calm, the air was very warm off the Gulf, the sky was clear, and a merciless sun began to generate a steaming inferno from the human and animal remains everywhere, especially within the debris wall sprawled across the city. Along the beachfront, most of the first two blocks were lost to the Gulf by wave erosion, and more than half of the built-up blocks of Galveston were swept clean (see fig. 1.6). Nearly all of the buildings behind the areas of total destruction were severely damaged (see fig. 1.7).

Because of the destruction of the bridges and cables, there was no immediate official transmission of the plight of the city to Houston and the rest of the nation on Sunday. But a local relief committee was organized to govern the city and begin to manage emergency operations, especially caring for survivors and disposing of the hundreds if not thousands of bodies (see fig. 1.8). The initial decision was to "bury" the weighted bodies

Fig. 1.5. Relief workers searching for bodies after the hurricane. Pictured is the leading edge of the wall of debris averaging six blocks across which was shoved on Galveston by the storm surge.

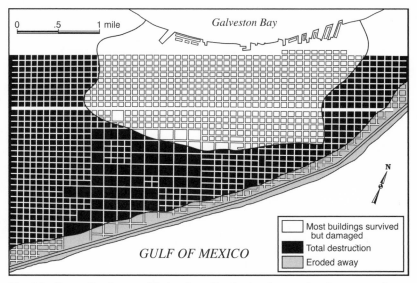

Fig. 1.6. Generalized map of Galveston after the hurricane, showing areas of erosion along the pre-storm beaches, areas of almost total destruction with few standing buildings, and the remaining areas with a mix of destroyed and heavily damaged buildings.

Fig. 1.7. Collage showing Galveston residential area where all buildings were down and destroyed (*top*); the debris wall near the center of the island (*middle*); the remaining walls of Sacred Heart Church (*lower left*); and a wooden home lifted off its foundation about ten blocks in from the beach (*lower right*).

Fig. 1.8. Removing dead bodies from the Galveston landscape for burial at sea.

at sea, and about seven hundred corpses were loaded on a barge late on Monday, and dumped in the sea Tuesday morning. By late Tuesday afternoon, however, some of the bodies began washing up on the beaches, and the relief committee terminated this program. In the humid tropical air decomposition was proceeding rapidly, and the relief committee was forced to order either immediate burial or burning. The funeral pyres persisted for several weeks, with the smoke and stench further increasing the misery of the survivors and relief workers.

A few survivors managed to cross Galveston Bay to the mainland on Sunday, but initially their stories of widespread destruction and horror were largely discounted as unsubstantiated wild rumors. It was not until Tuesday morning that the first relief train from Houston was staffed and outfitted with supplies for Galveston. The train was not able to reach the point where the railroad trestle had crossed Galveston Bay. The relief team along with some Galveston residents had to first walk several miles along the railroad tracks across the coastal marshes and then commandeer an abandoned vessel for the crossing. The bay was so filled with floating debris and bodies that the crossing was not completed until Tuesday evening, and the team first set foot at the harbor in Galveston Wednesday morning, almost seventy-two hours after the storm had abated.

Altogether between six and twelve thousand people lost their lives in Galveston and the adjacent lowlands of southeastern Texas, making this storm the greatest single natural disaster in terms of loss of life in the history of the United States to this very day. More than a third of the buildings in Galveston were destroyed, and most of the remainder were heavily damaged. Entrepreneurs in Houston were able to seize this opportunity to plan and develop the Houston Ship Canal as a safer and much more efficient location for port facilities, and Galveston was never able to reclaim its status as the leading American port on the western Gulf.

Nevertheless, the city developed an engineering plan not only to re- build but to make the city much safer from hurricane destruction. A granite boulder seawall, seventeen feet high and twenty-seven feet wide, and originally three miles long, was built along the beachfront (see fig. 1.9), with a promenade and road on top, and later extended in 1962 to ten miles in length (see fig. 1.10). In addition, in a colossal engineering feat for its time, the elevation of the entire city was raised block by block with fill from the sandy bottom of Galveston Bay (see figs. 1.11 and 1.12), with over two thousand buildings including a magnificent cathedral jacked up to the new levels (see fig. 1.13). All of this work was completed by 1910 and helped to prevent widespread damage by major hurricanes, cat- egory three or greater, in 1915, 1932, 1949, and 1983 (Hurricane Alicia).

Fig. 1.9. Galveston's 3-mile, 17-foot sea wall and promenade were completed by 1904, as part of a comprehensive plan to secure the city against future hurricanes.

Fig. 1.10. The Galveston sea wall in more modern times.

Fig. 1.11. Sand and water slurry from Galveston Bay were pumped through pipes into sectors of the city to raise land elevations within the city.

Fig. 1.12. Galveston residences and the surrounding land were raised to provide protection from future storm surges.

Fig. 1.13. The 3,000-ton St. Patrick's Church was raised five feet with 700 jacks.

Although the city lost its primary function as a seaport, it gradually transformed into a beach resort for Texas and much of the Great Plains.

There were a number of deadly hurricanes along the Gulf Coast in the twentieth century, but none caused such catastrophic loss of life and property as that seen in Galveston in September 1900. Much of the reduction in fatalities can be attributed to improvements in storm forecasting and emergency management. In 2005, however, more than one hundred years later, Hurricane Katrina came on shore, first in Florida and then in southeastern Louisiana and Mississippi. Katrina serves as a grim reminder that a century later our ability to plan for cataclysmic storms is still far less than satisfactory.

| 2 |

Hurricane Katrina

The Galveston Hurricane of 1900 was the most devastating hurricane along the American coast in recorded history in terms of loss of life. For Weather Bureau forecasters of that time, hurricane tracking was not much more than an experienced guessing game. Identification of storms mostly came from previous landfalls. Forecasting was also crude and largely dependent on prior tracks over the West Indies, the Florida Peninsula, and the Yucatan, as well as coastal observations of wind speed and direction, and atmospheric pressure changes. Furthermore, there was no available technology for wireless reports from ships caught off guard in the path of storms. Hence, the oceans and seas were blank on the simple weather maps of the time.

Hurricane Katrina, which hit the Gulf Coast in 2005, was the most destructive hurricane in terms of property loss along the American coast in recorded history. It made landfall along the northern Gulf Coast a little more than three hundred miles from (and 105 years after) the Galveston Hurricane, but the technologies for monitoring and predicting storm tracks and intensities and systems for search and rescue were ages and worlds apart. By 2005, tropical storms and hurricanes could not escape detection by satellites, coastal radar, and weather buoys, and real-time radio reports came in from ships and planes. The real-time data were supplemented by sophisticated computer models utilized for prediction of future tracks, intensities, and landfalls by a dedicated federal office—the National Hurricane Center in Miami, a unit of the National Weather Service and the National Oceanic and Atmospheric Administration, better known by its acronym, NOAA. Nevertheless, we will see that uncertainties about the predicted track and behavior of Katrina, up until about sixty hours before landfall in southeast Louisiana and coastal Mississippi, delayed residential evacuations and inhibited mitigation efforts by public and private agencies both before and after the storm.

One hundred years after Galveston, hurricane strikes were potentially much more hazardous because of the denser population of coastal cities and beaches. For the federal census of 1900, there were only about 38,000 inhabitants in Galveston and an additional 6,000 people outside of the city on the barrier island; as much as 20 to 25 percent of this population lost their lives during or shortly after the hurricane. In 2005, in contrast, more than 1.6 million people resided in the Katrina-affected parishes in southeastern Louisiana (Plaquemines, St. Bernard, Orleans, Jefferson, Lafourche, and St. Tammany) and the three coastal counties of Mississippi (Hancock, Harrison, and Jackson). Hence, in 2005 the federal, state, and local officials responsible for the safety and survival of their constituents had to cope with far more complex human and economic situations than did their counterparts in 1900, despite all the technology in their arsenal.

Scientists who specialize in hurricanes and natural hazards have regarded New Orleans as one of the greatest American natural disasters waiting to happen; it was not "if" but rather "when." These dire predictions were not based solely on the frequency of hurricane strikes along the central Gulf Coast but also on the unique geographical setting of the city and metropolitan region (see fig. 2.1). In the early nineteenth century, New Orleans occupied the low but mostly well drained natural levee lands along the banks of the Mississippi River. As the city expanded in the later nineteenth and early twentieth centuries, the extensive marshlands back from the river were drained for urban developments, surrounding levees were erected to keep out river and Gulf waters, and massive pumps were installed to drain rainwater over the levees. The marsh soils compacted as they were dewatered, and regional subsidence associated with oil and gas extraction as well as the absence of river sediments, usually distributed across the delta by floods but now kept out by the levees, resulted in many base elevations within the levee systems falling below sea level. By 1994 the New Orleans Planning Commission estimated that only 45 percent of the developed area of the city was above sea level, with neighborhoods ranging from twenty feet above sea level along the Mississippi River to more than ten feet below, especially in eastern New Orleans, with the very lowest locations more than fifteen feet below sea level (see fig. 2.2). If the levees were ever breached, the city could fill rapidly with a potentially toxic brew of sea and river water mixed with urban liquids and solids (oil, gas, paint thinner, antifreeze, etc.) that would remain in the city for weeks until all could be pumped back out again. Despite these concerns about this great natural disaster waiting to happen, effective evacuation plans at local, state, and federal levels remained incomplete and not well coordinated.

It is also important to appreciate that 2005 was by far the most active tropical storm and hurricane season over the entire Atlantic Basin in the archives and official records of the National Hurricane Center dating back to the middle nineteenth century. During the 2005 season, there was a record-breaking total of twenty-seven named storms, with the previous record of twenty-one storms in 1933, and a forty-year average of only eleven storms per year. There were fifteen hurricanes during 2005, as opposed to the previous record of twelve in 1969 and an annual average of six hurricanes. There were four category five storms—Emily, Katrina, Rita, and Wilma—and all but Emily attained this maximum severe status over the Gulf of Mexico. Hurricane Wilma set a new record mini-

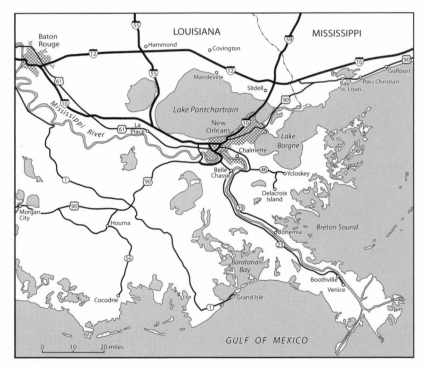

Fig. 2.1. Regional setting of southeastern Louisiana and coastal Mississippi.

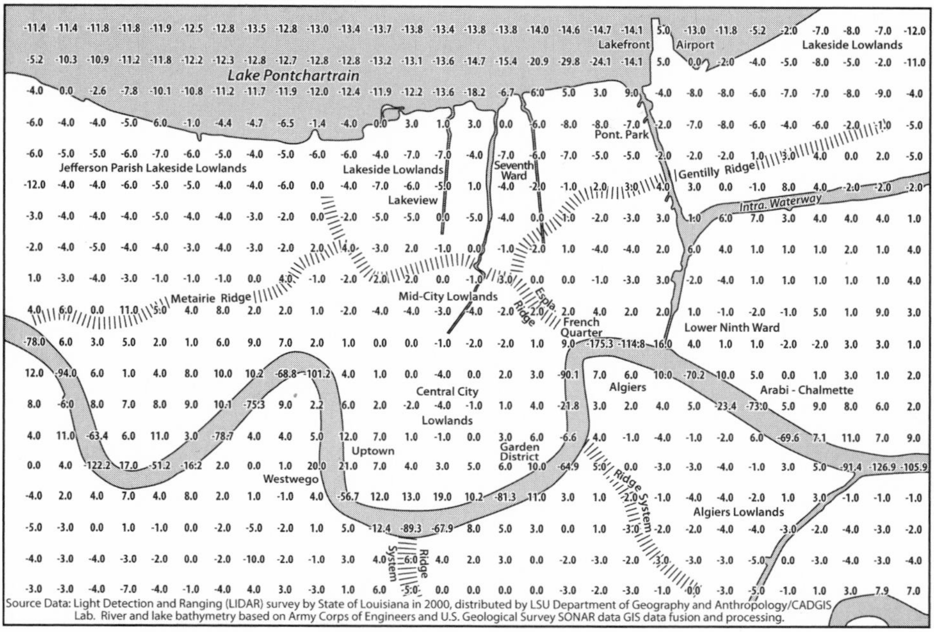

-11.4	-11.4	-11.8	-11.8	-11.9	-12.5	-12.8	-13.5	-12.8	-13.0	-13.4	-13.7	-13.8	-13.4	-13.8	-13.8	-14.0	-14.6	-14.7	-14.1	5.0	-13.0	-11.8	-5.2	-2.0	-7.0	-8.0	-7.0	-12.0

Lakefront Airport — Lakeside Lowlands

| -5.2 | -10.3 | -10.9 | -11.2 | -11.8 | -12.2 | -12.3 | -12.8 | -12.7 | -12.8 | -12.8 | -13.2 | -13.1 | -13.6 | -14.7 | -15.4 | -20.9 | -29.8 | -24.1 | -14.1 | 5.0 | 0.0 | -2.0 | -4.0 | -5.0 | -8.0 | -5.0 | -2.0 | -11.0 |

Lake Pontchartrain

| -4.0 | 0.0 | -2.6 | -7.8 | -10.1 | -10.8 | -11.2 | -11.7 | -11.9 | -12.0 | -12.4 | -11.9 | -12.2 | -13.6 | -18.2 | -6.7 | -6.0 | 5.0 | 3.0 | 9.0 | -4.0 | -8.0 | -8.0 | -6.0 | -7.0 | -7.0 | -8.0 | -9.0 | -4.0 |

| -6.0 | -4.0 | -4.0 | -5.0 | -6.0 | -1.0 | -4.4 | -4.7 | -6.5 | -1.4 | -4.0 | -0.0 | 3.0 | 1.0 | 3.0 | 0.0 | 6.0 | -8.0 | -8.0 | -7.0 | -2.0 | -7.0 | -8.0 | -6.0 | -4.0 | -6.0 | -2.0 | 1.0 | -5.0 |

Pont. Park

| -6.0 | -5.0 | -5.0 | -6.0 | -7.0 | -6.0 | -5.0 | -4.0 | -5.0 | -6.0 | -6.0 | -4.0 | -7.0 | -7.0 | -4.0 | -7.0 | -6.0 | -7.0 | -5.0 | -5.0 | -2.0 | -2.0 | -2.0 | 1.0 | 9.0 | 4.0 | 0.0 | 2.0 | -5.0 |

Jefferson Parish Lakeside Lowlands — Lakeside Lowlands — Seventh Ward — Gentilly Ridge

| -12.0 | -4.0 | -4.0 | -6.0 | -5.0 | -5.0 | -4.0 | -4.0 | -6.0 | 0.0 | -4.0 | -7.0 | -6.0 | -5.0 | 1.0 | -4.0 | -2.0 | -1.0 | 2.0 | 3.0 | 4.0 | 3.0 | 0.0 | -1.0 | 8.0 | 4.0 | -2.0 | -2.0 | -2.0 |

Lakeview — Intra. Waterway

| -3.0 | -4.0 | -4.0 | -4.0 | -5.0 | -4.0 | -4.0 | -3.0 | -2.0 | 0.0 | -2.0 | -5.0 | -5.0 | 0.0 | -5.0 | -4.0 | 0.0 | 0.0 | -2.0 | -3.0 | 3.0 | 1.0 | -6.0 | 7.0 | 3.0 | 4.0 | 4.0 | 4.0 | 1.0 |

| -2.0 | -4.0 | -5.0 | -4.0 | -4.0 | -3.0 | -4.0 | -3.0 | -2.0 | 2.0 | 4.0 | -3.0 | 2.0 | 1.0 | 0.0 | -1.0 | -2.0 | 1.0 | -4.0 | -4.0 | 2.0 | 6.0 | 4.0 | 1.0 | 1.0 | 1.0 | 2.0 | 1.0 | 4.0 |

Mid-City Lowlands

| 1.0 | -3.0 | -4.0 | -3.0 | -1.0 | -1.0 | -1.0 | 0.0 | 4.0 | -1.0 | -2.0 | -2.0 | 2.0 | 0.0 | -1.0 | 3.0 | 0.0 | -1.0 | -3.0 | 3.0 | -2.0 | -4.0 | 1.0 | 1.0 | 1.0 | 1.0 | 1.0 | 4.0 |

Metairie Ridge

| 4.0 | 6.0 | 0.0 | 11.0 | 5.0 | 4.0 | 8.0 | 2.0 | 2.0 | 1.0 | -2.0 | -4.0 | -4.0 | -3.0 | 4.0 | -2.0 | 2.0 | 4.0 | 2.0 | 2.0 | 1.0 | -1.0 | -2.0 | -1.0 | 5.0 | 1.0 | 9.0 | 3.0 |

French Quarter — Lower Ninth Ward

| -78.0 | 6.0 | 3.0 | 5.0 | 2.0 | 1.0 | 6.0 | 9.0 | 7.0 | 2.0 | 1.0 | 0.0 | 0.0 | -1.0 | -2.0 | -2.0 | 1.0 | 9.0 | -175.3 | -114.8 | -16.0 | 4.0 | 1.0 | 1.0 | -2.0 | -2.0 | 3.0 | 3.0 | 1.0 |

| 12.0 | -94.0 | 6.0 | 1.0 | 4.0 | 8.0 | 10.0 | 10.2 | -68.8 | -101.2 | 4.0 | 1.0 | 0.0 | -4.0 | 0.0 | 2.0 | 3.0 | -90.1 | 7.0 | 6.0 | 10.0 | -70.2 | -10.0 | 5.0 | 0.0 | 1.0 | 3.0 | 1.0 | 2.0 |

Central City — Algiers — Arabi - Chalmette

| 8.0 | -6.0 | 8.0 | 7.0 | 8.0 | 9.0 | 10.1 | -75.3 | 9.0 | 2.2 | 6.0 | 2.0 | -4.0 | -1.0 | 1.0 | 4.0 | -21.8 | 2.0 | 4.0 | 5.0 | -23.4 | -73.0 | 5.0 | 9.0 | 8.0 | 6.0 | 2.0 |

Lowlands

| 4.0 | 11.0 | -63.4 | 6.0 | 11.0 | 3.0 | -78.7 | 4.0 | 4.0 | 5.0 | 12.0 | 7.0 | 1.0 | -1.0 | 0.0 | 3.0 | 6.0 | -6.6 | 4.0 | -1.0 | -4.0 | -1.0 | -2.0 | 6.0 | -69.6 | 7.1 | 11.0 | 7.0 | 9.0 |

Uptown — Garden District

| 0.0 | 4.0 | -122.2 | -17.0 | -51.2 | -16.2 | 2.0 | 0.0 | 1.0 | 20.0 | 21.0 | 7.0 | 4.0 | 3.0 | 5.0 | 6.0 | 10.0 | -64.9 | 5.0 | 0.0 | -3.0 | -3.0 | -4.0 | 3.0 | 5.0 | -91.4 | -126.9 | -105.9 |

Westwego

| -4.0 | 2.0 | 4.0 | 7.0 | 4.0 | 8.0 | 2.0 | 1.0 | -1.0 | 4.0 | -56.7 | 12.0 | 13.0 | 19.0 | 10.2 | -81.3 | 11.0 | 3.0 | 1.0 | -2.0 | -1.0 | -4.0 | -4.0 | -2.0 | 1.0 | 3.0 | -1.0 | -1.0 | -1.0 |

Algiers Lowlands

| -5.0 | -3.0 | 0.0 | 1.0 | -1.0 | 0.0 | -2.0 | -5.0 | -2.0 | 1.0 | 5.0 | -12.4 | -89.3 | -67.9 | 8.0 | 5.0 | 3.0 | 0.0 | 1.0 | -3.0 | -2.0 | -2.0 | -4.0 | -4.0 | -3.0 | -2.0 | -4.0 | -3.0 | -2.0 |

| -4.0 | -3.0 | -4.0 | -3.0 | -2.0 | 0.0 | -2.0 | -10.0 | -2.0 | -1.0 | 3.0 | 4.0 | -6.0 | 4.0 | 2.0 | 3.0 | 0.0 | -2.0 | -3.0 | -2.0 | 3.0 | -3.0 | -3.0 | -5.0 | 0.0 | -3.0 | -4.0 | -4.0 |

| -3.0 | -3.0 | -4.0 | -7.0 | -4.0 | -1.0 | -4.0 | 4.0 | 3.0 | -3.0 | 1.0 | 6.0 | -5.0 | 0.0 | 0.0 | 0.0 | 0.0 | -3.0 | -2.0 | -3.0 | -1.0 | -0.0 | -3.0 | -5.0 | -1.0 | 1.0 | 3.0 | 7.9 | 7.0 |

Source Data: Light Detection and Ranging (LIDAR) survey by State of Louisiana in 2000, distributed by LSU Department of Geography and Anthropology/CADGIS Lab. River and lake bathymetry based on Army Corps of Engineers and U.S. Geological Survey SONAR data GIS data fusion and processing.

Fig. 2.2. Gridded elevation values across New Orleans, showing areas above and below sea level.

mum atmospheric pressure for the Atlantic Basin (see table 2.1 for Saffir-Simpson Hurricane Intensity Scale).

During the early 2005 season, there were many days when tropical storms and hurricanes threatened life and property around the shores of the Gulf of Mexico. By mid-August a record-breaking four tropical storms and three hurricanes had been somewhere over the Gulf for a total equivalent of nine days of storms, with two storms, Dennis and Emily, attaining major status. Emily, along with Tropical Storms Bret, Gert, and José, went to Mexico. Prior to Katrina, Tropical Storm Arlene came ashore near Pensacola Beach, and major Hurricane Dennis further damaged the beach communities along the Florida Panhandle, ravaged a year earlier by major Hurricane Ivan. In between, Hurricane Cindy made landfall as a tropical storm in the vicinity of Grand Isle, Louisiana, and passed directly over the New Orleans metropolitan area, causing only moderate damage. So when the tropical disturbance later called Katrina was first identified by the National Hurricane Center (NHC) on August 23 over the southern Bahamas, residents of New Orleans and the northern Gulf Coast could either become anxious again or think "another false alarm, why be concerned?"

Table 2.1. Saffir-Simpson Hurricane Intensity Scale

Scale Number (category)	Winds (mph)	Typical Characteristics of Hurricanes by Category			
		Central Pressure		Surge (feet)	Damage
		(millibars)	(inches, inHg)		
1	74–95	> 979	> 28.91	4 to 5	Minimal
2	96–110	965–979	28.50–28.91	6 to 8	Moderate
3	111–130	945–964	27.91–28.47	9 to 12	Extensive
4	131–155	920–944	27.17–27.88	13 to 18	Extreme
5	> 155	< 920	< 27.17	> 18	Catastrophic

What follows is an account of the development and forecasts of Katrina as it formed as a tropical depression (TD) over the southern Bahamas about one thousand miles southeast of New Orleans. It then crossed southern Florida, powered up to a category five over the central Gulf, and came on shore as a category three just south of New Orleans near Buras, Louisiana, early on Monday, August 29. We simplified, shortened, and omitted some advisories from the NHC, and the timeline for Louisiana landfall is our own.

Tuesday, August 23, 5 p.m. EDT—TD 12: Birth over the Southern Bahamas—5 days and 12 hours before landfall in Louisiana

In NHC Advisory Number 1 at 5 p.m. Eastern Daylight Time (EDT), the National Hurricane Center officially designated disturbed tropical weather as TD 12, positioned about 175 miles south of Nassau and moving slowly toward the northwest at 8 miles per hour. Top winds were estimated near 35 miles per hour, and the lowest atmospheric pressure was 29.74 inches (1007 millibars). TD 12 was expected to gradually intensify, and the Bahamian Weather Service issued tropical storm warnings for the central and northwest Bahamas. The NHC projected the depression to develop to tropical storm (TS) status as TS Katrina, crossing Florida and reaching the northeastern Gulf of Mexico off Cedar Key, Florida, by

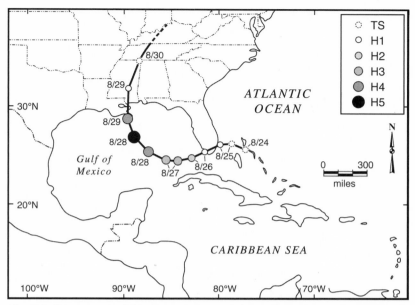

Fig. 2.3. Storm track of Hurricane Katrina as depicted by the National Hurricane Center.

Saturday, August 27. Figure 2.3 shows the storm track from the Bahamas to Louisiana and Saffir-Simpson intensity categories along the way.

Tuesday, August 23, 11 p.m. EDT—TD 12: Moving slowly toward Florida—5 days and 6 hours before landfall in Louisiana

Advisory 2 indicated no changes in direction or intensity with TD 12 about 140 miles southeast of Nassau. Nevertheless, the forward progress of TD 12 resulted in the NHC posting tropical storm watches along the eastern coast of Florida from Vero Beach southward to the middle Florida Keys. Expectations at the NHC continued to focus on a tropical storm emerging over the northeastern Gulf in the vicinity of Cedar Key, and then turning northward toward the eastern Florida Panhandle.

Wednesday, August 24, 5 a.m. EDT—TD 12: Closer to Florida—about 5 days before landfall in Louisiana

Advisory 3 continued to show that TD 12 was closing in on a strike along Florida's east coast. Intensity and track predictions remained unchanged, with pressure falling slightly to 29.71 inches (1006 millibars). The center was located about 95 miles southeast of Nassau, or about 270 miles southeast of southern Florida. Tropical storm watches were not changed, and projections continued to indicate a tropical storm over the northeastern Gulf by Saturday.

Wednesday, August 24, 5 p.m. EDT—TS Katrina: Which way after reaching the Gulf of Mexico?—4 days and 12 hours before landfall in Louisiana

Advisory 5 indicated that TD 12 had strengthened to tropical storm status and was now named TS Katrina. Pressure had fallen to 29.59 inches (1002 millibars) with sustained winds at 45 miles per hour. Katrina was located about 45 miles north-northeast of Nassau, about 195 miles east of the Florida Coast, with forward movement at 9 miles per hour toward the northwest. Tropical storm warnings were posted from Vero Beach southward to Florida City and tropical storm watches both northward and southward beyond the TS warnings.

The various meteorological computer models utilized by the NHC at this time were not in agreement about the future routing of TS Katrina. An upper-air east-west ridge across northern Florida and the southern states was expected to steer Katrina westward out over the Gulf, but several of the models predicted a weakening of the ridge over the northeastern Gulf and an opportunity for Katrina to turn sharply northward. The most likely track brought Katrina as a hurricane northward to the eastern Florida Panhandle between St. Marks and Apalachicola Saturday afternoon, but the NHC did allow for a 2 percent chance of landfall as far west as New Orleans. Even though Katrina was still off the east coast of Florida, this was the first official hint of a potential threat to southern Mississippi and southeastern Louisiana, four and a half days prior to actual landfall.

Wednesday, August 24, 11 p.m. EDT—TS Katrina: A slight turn toward the west—18 hours before landfall in Florida and 4 days and 6 hours before landfall in Louisiana

Advisory 6 reported that TS Katrina had turned ever so slightly toward the west with little change of forward speed estimated at 8 miles per hour. The center was reported to be 60 miles southeast of Freeport in the Bahamas, or about 135 miles east of Miami on the Florida Coast. Central pressure had fallen a little to 29.56 inches (1001 millibars), and sustained winds had increased to 50 miles per hour with higher gusts. Hurricane warnings were posted by the NHC from Vero Beach south to Florida City, 30 miles southwest of Miami, placing almost 150 miles of a very densely populated coast and beaches within the warning sector. Tropical storm watches were extended an additional 70 miles to the north to Titusville and about 50 miles southwest in the Florida Keys to Marathon. After reaching the Gulf, Hurricane Katrina was still expected to curve sharply to the north and come on shore Saturday evening between St. Marks and Apalachicola. Nevertheless, the probability of a strike at New Orleans was raised from 2 to 3 percent.

Thursday, August 25, 5 a.m. EDT—TS Katrina: No significant changes—12 hours before landfall in Florida and 4 days before landfall in Louisiana

Advisory 7 placed TS Katrina 90 miles east of Fort Lauderdale with continued movement to the west at 8 miles per hour. Pressure was down a bit to 29.53 inches (1000 millibars), but no increase of maximum winds at 50 miles per hour was reported. The extent of hurricane warnings and tropical storm watches remained unchanged, but the highest probability for landfall on the northern Gulf Coast was shifted slightly to the west, now between Panama City and St. Marks, and the probability of a later strike at New Orleans was raised a notch to 4 percent.

Thursday, August 25, 5 p.m. EDT—Hurricane Katrina coming ashore near Fort Lauderdale—3 days and 12 hours before landfall in Louisiana

Advisory 9 placed the center of now category one Hurricane Katrina 15 miles east-northeast of Fort Lauderdale, still moving west but a bit slower at 6 miles per hour. Sustained winds had increased to 75 miles per hour with higher gusts, and the central pressure had fallen rapidly to 29.09 inches (985 millibars). Tropical storm warnings were hoisted for the Florida Keys and northward along the west coast to Longboat Key, a barrier island northwest of Sarasota. Projections for landfall of Hurricane Katrina in the Florida Panhandle were again shifted to the west, with the most likely landfall estimated to be near Apalachicola (see fig. 2.4). Some of the computer models suggested a strike as far west as New Orleans, and the official probability of a strike there was raised to 7 percent.

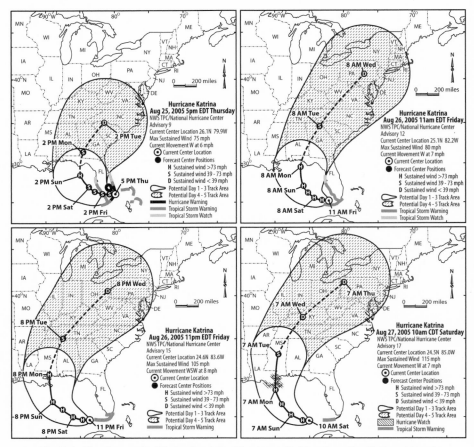

Fig. 2.4. The forecasted 5-day track for Hurricane Katrina from the National Hurricane Center: (*upper left*) 5 p.m. EDT, Thursday, Aug. 25; (*upper right*) 11 a.m. EDT, Friday, Aug. 26; (*lower left*) 11 p.m. EDT, Friday, Aug. 26; (*lower right*) potential 3-day track, 10 p.m. CDT, Saturday, Aug. 27.

Thursday, August 25, 11 p.m. EDT: Hurricane Katrina moving to the southwest across southern Florida—3 days and 6 hours before landfall in Louisiana

Advisory 10 reported that Hurricane Katrina was moving southwestward at 8 miles per hour with the center on shore about 20 miles northwest of Homestead Florida. Sustained winds were estimated at 75 miles per hour and the central pressure had fallen slightly to 29.06 inches (984 millibars) in spite of landfall and the short passage over mostly low-lying wetlands and swamps. The storm was expected to reach the Gulf of Mexico at minimal hurricane status and turn gradually toward the north, with the most probable landfall in the vicinity of Pensacola on Sunday afternoon. The probability of a hurricane strike at New Orleans was maintained at 7 percent.

Friday, August 26, 5 a.m. EDT: Hurricane Katrina moving westward across the central Gulf—3 days before landfall in Louisiana

Advisory 11 indicated that Hurricane Katrina was positioned about 50 miles north-northeast of Key West with slower forward motion of only 5 miles per hour but turning more westerly than southwest. Sustained winds were estimated at minimal hurricane strength of 75 miles per hour, and the central pressure had risen a little to 29.15 inches (987 millibars). A gradual increase of intensity was predicted as the hurricane was expected to turn gradually toward the northwest and north, but the probability of a strike at New Orleans was raised to 8 percent.

In the morning edition of the *New Orleans Times-Picayune*, prepared Thursday evening, the Katrina story was set back on page A-10, with a summary of storm impacts in Florida and the expectation that Katrina would turn northward and strike the eastern Florida Panhandle. There was hardly a hint of potential danger to New Orleans and southeastern Louisiana.

Friday, August 26, 11 a.m. EDT: Category 1 Hurricane Katrina moving westward—only 2 days and 18 hours before landfall in Louisiana

Advisory 12 had Katrina located 45 miles northwest of Key West and moving westward at 7 miles per hour as a minimal category one storm with peak winds of 80 miles per hour. Pressure was estimated at 28.97 inches (981 millibars) and falling. Two of the prediction models utilized by the NHC predicted landfall in Louisiana, but the majority still called for landfall in the eastern Florida Panhandle (see fig. 2.4). Advisory 13 was issued half an hour later at 11:30 a.m. EDT reporting that the latest data from reconnaissance aircraft indicated significant intensification to category two, with pressure down to 28.67 inches (971 millibars).

Friday, August 26, 5 p.m. EDT: Category 2 Hurricane Katrina strengthening over central Gulf—only 2 days and 12 hours before landfall in Louisiana

The NHC reported in Advisory 14 that Katrina was moving west-southwest at 8 miles per hour, 70 miles west-northwest of Key West. Pressure had fallen significantly to 28.50 inches (965 millibars), and sustained winds had increased to 100 miles per hour with higher gusts. Katrina was now a category two hurricane and was projected to move over the very warm "loop current" for a day or more, with the prediction of intensifying to category three shortly, and perhaps to category four on Saturday. Expectations now were that Katrina would strike southeastern Louisiana Monday afternoon somewhere between Gulfport, Mississippi, to the east and as far as New Iberia to the west, with New Orleans in the center of the projection.

Friday, August 26, 11 p.m. EDT: Category 2 Katrina continues on southwesterly track but threatens New Orleans and southeast Louisiana—landfall only 2 days and 6 hours ahead

In Advisory 15 the NHC reported that movement continued toward the west-southwest at 8 miles per hour, at a location 460 miles southeast of the mouth of the Mississippi River. The central pressure remained at 28.50 inches (965 millibars), but sustained winds had increased slightly to 105 miles per hour. The computer models suggested a possible track and landfall in the vicinity of New Iberia, Louisiana, almost 100 miles west of New Orleans, on Monday evening, but some of the models suggested even more curvature to the north and a landfall strike in southern Mississippi (see fig. 2.4). Nevertheless, the official consensus prediction continued to be Plaquemines Parish and New Orleans.

Saturday, August 27, 5 a.m. EDT: Katrina intensifies to Category 3—only two days from landfall in southeastern Louisiana

The NHC reported in Advisory 16 that Katrina was turning slowly to the right and moving toward the west at 7 miles per hour, and located about 430 miles southeast of the lower delta of the Mississippi River. Pressure had fallen rapidly to 27.91 inches (945 millibars), and sustained winds had increased to 115 miles per hour with higher gusts. Hurricane-force winds extended out ahead of the center for 40 miles and gale-force winds, 150 miles. The NHC shifted the projected track at landfall somewhat back to the east, with landfall in the lower delta of the Mississippi River south of New Orleans in Plaquemines Parish early Monday morning, probably as a category four storm. The *Times-Picayune* finally placed Katrina as the lead story on page A-1, with appropriate warnings of the threat to the city and surrounding region as well as the Mississippi Gulf Coast.

Saturday, August 27, 10 a.m. CDT: Katrina is an increasing threat for New Orleans—less than 48 hours before landfall

In Advisory 17 the NHC positioned category three Katrina at 405 miles southeast of the lower delta of the Mississippi River, with the storm still moving west at 7 miles per hour. Sustained winds were maintained at 115 miles per hour, but the central pressure continued to fall to 27.76 inches (940 millibars). Katrina could well be a category four at landfall, which was expected in less than 48 hours. The various computer models predicted landfall from New Iberia eastward almost to Mobile, Alabama, but the focus clearly remained on southeast Louisiana, New Orleans, and southern Mississippi.

At 9 a.m. two of the parishes in the vicinity of New Orleans, Plaquemines and St. Charles, had ordered mandatory evacuations, and at 10 a.m., the NHC issued a hurricane watch for metropolitan New Orleans.

Saturday, August 27, 4 p.m. CDT: Still closer, but not much change in track or intensity—less than 40 hours before landfall

Advisory 18 indicated little change of the track and intensity of Katrina since the previous advisory. Katrina was now 380 miles southeast of the Lower Delta, moving west at 7 miles per hour, with the pressure up slightly to 27.91 inches (945 millibars), and sustained winds of 115 miles per hour. Hurricane winds extended 45 miles out from the eye, and tropical storm–force winds 160 miles. A hurricane watch was posted from Intracoastal City, Louisiana, eastward more than 200 miles to the Mississippi-Alabama border.

At 5 p.m. Mayor Ray Nagin finally declared a state of emergency in New Orleans, and at the same time he issued a voluntary evacuation order for the city. Mandatory evacuation orders had already been issued for low-lying areas of the parishes downstream from New Orleans, along the shores of Lake Pontchartrain, and for much of the Mississippi and Alabama Gulf Coast. However, some residents were reluctant to leave for various reasons, among them the recent memories of dreadful experiences along evacuation routes out of the city for Hurricane Georges in 1998 and Hurricane Ivan just the year before in 2004, when, for example, the 60- to 90-minute drive from New Orleans to Baton Rouge evolved into an endurance event of up to 12 hours for many evacuees. Both Georges and Ivan during the final hours before landfall shifted to the east, resulting in only relatively minor damage in New Orleans. In addition, Katrina was still not moving directly toward New Orleans, with some hoping that this storm would again bypass New Orleans and this time continue westward toward the Texas coast.

Saturday, August 27, 10 p.m. CDT: A slight turn toward the northwest—less than 36 hours before landfall

According to Advisory 19, Hurricane Katrina was now about 335 miles south-southeast of the Lower Delta, and moving west-northwest at 7 miles per hour. Pressure had fallen again to 27.73 inches (939 millibars), with sustained winds at 115 miles per hour. A further turn to the right was anticipated, with hurricane warnings from Morgan City, Louisiana, eastward to the Alabama-Florida border, a distance of about 175 miles. Hurricane watches and tropical storm warnings were issued for Morgan City west to Intracoastal City and for the Florida Panhandle eastward to Destin. The most likely strike remained the New Orleans metropolitan region on Monday morning (see fig. 2.4). The director of the NHC, Max Mayfield, personally telephoned officials in Louisiana, Mississippi, and Alabama, warning of the potential of extreme hurricane conditions.

Sunday, August 28, 4 a.m. CDT: Katrina intensifies rapidly to a category 4 and moves more quickly toward New Orleans—little more than 24 hours before landfall

Advisory 21 located Hurricane Katrina about 275 miles south-southeast of the Lower Delta, with increasing forward speed of 10 miles per hour toward the west-northwest. Central pressure had fallen to 27.61 inches (935 millibars), and sustained winds had increased to 145 miles per hour with higher gusts. Hurricane-force winds now extended out 85 miles ahead of the eye, and tropical storm–force winds 185 miles. Because Katrina had become a much larger storm, tropical storm warnings were extended west to Cameron, Louisiana, and east to Indian Pass, Florida.

Mayor Nagin finally issued a mandatory evacuation order for New Orleans at 9:30 a.m. and predicted landfall little more than 24 hours away, with the Superdome designated as the shelter of last resort.

Sunday, August 28, 10 a.m. CDT: Katrina "explodes" to Category 5 and moves even more quickly toward New Orleans—less than 24 hours before landfall

Advisory 23 placed category five Hurricane Katrina 225 miles south-southeast of the Lower Delta, moving west-northwest at 12 miles per hour. The central pressure had "crashed" to 26.78 inches (907 millibars), and sustained winds had increased to 175 miles per hour with higher gusts. Hurricane-force winds now reached out 105 miles ahead of the eye, and tropical storm winds, 205 miles. Hence, tropical storm–force winds would reach Plaquemines Parish, the bird's foot delta of the Mississippi River below New Orleans, in less than three hours. The geographical extent of hurricane and tropical storm warnings remained essentially unchanged since Advisory 21. The NWS warned again that residents of southeast Louisiana and southern Mississippi could expect a truly catastrophic event.

Sunday, August 28, 4 p.m. CDT: Very stormy conditions begin to sweep over Plaquemines Parish even though Katrina's eye is still 150 miles south of the bird's foot delta—less than 18 hours before landfall

The eye of category five Katrina was located about 150 miles south of Plaquemines Parish, moving northwest with increasing forward speed of 13 miles per hour, according to Advisory 24 by the NHC. The central pressure had fallen to 26.64 inches (902 millibars), but sustained winds had slowed just a bit to 165 miles per hour.

At this time it was estimated that about 80 percent of the residents of metropolitan New Orleans had evacuated to the north and west, but about 100,000 people still remained in the area, many without vehicles or financial resources to leave and relocate away from the threatened city.

Sunday, August 28, 10 p.m. CDT: Katrina begins to lash all of southeastern Louisiana and southern Mississippi

Advisory number 25 from the NHC placed category five Katrina 105 miles south of Plaquemines Parish and 170 miles south-southeast of New Orleans. Katrina was moving toward the north-northwest at 10 miles per hour with sustained winds of 160 miles per hour. The central pressure had risen slightly to 26.70 inches (904 millibars). Outer torrential rain bands and damaging winds were beginning to sweep over the city and the Mississippi Gulf Coast. Between 8,000 and 10,000 people were already crowded into the Superdome, the shelter of last resort in New Orleans. Tens of thousands of people remained in harm's way below sea level in New Orleans and within several miles of the beaches along the Mississippi Coast. Katrina's projected track would take it very close to New Orleans with landfall in coastal Mississippi in the vicinity of Bay St. Louis, where Hurricane Camille came on shore as a category five storm in 1969 with overwhelming devastation. Among those who did not respond to evacuation orders before Hurricane Camille, 100 to 200 people lost their lives from Bay St. Louis eastward to Biloxi. As Katrina approached, the few evacuation highways out of New Orleans quickly flooded and became impassible. Although there were more roads and highways in coastal Mississippi that provided opportunities to get away to the north ahead of the storm, most of the bridges over the coastal waterways and sounds were destroyed during Katrina, isolating the coastal towns and cities from the north for several days.

Monday, August 29, 4 a.m. CDT: Hurricane conditions over southeastern Louisiana and coastal Mississippi

Advisory 26 of the NHC reported that the center of Katrina was located about 90 miles south-southeast of New Orleans and 120 miles south-southwest of Biloxi. Category four Katrina was moving northward at 15 miles per hour with sustained winds of 150 miles per hour. Katrina was beginning to weaken because the outer circulation to the north was over land and drier air was beginning to enter the system from the west.

Landfall occurred about 6 a.m. near Buras in Plaquemines Parish, 50 miles southeast of New Orleans, as a category three storm. Unfortunately, Katrina had been a category five just six hours before landfall and still generated a record storm surge that destroyed nearly every dwelling in middle and lower Plaquemines Parish. At 4 a.m. extremely destructive winds, tornadoes, near-record storm surges, and heavy rains were projected to ravage New Orleans and coastal Mississippi for the remainder of the day. The track of the center was now predicted to pass just east of New Orleans, keeping most of the city in the less intense left side of the storm, with mostly northerly rather than southerly winds. But this same track would intensify winds and storm surges along most of coastal Mississippi.

In New Orleans the storm surge coming in from the east began to generate leaks from the Industrial Canal near the I-10 bridge, causing minor flooding in the Gentilly and New Orleans East neighborhoods. By 5 a.m. the levees along the Mississippi River-Gulf Outlet Canal (Mr. Go) began to crumble, allowing the storm surge to approach Chalmette in St. Bernard Parish from the north and east. Water rose so quickly around 8 a.m. that some remaining residents were unable to escape; much media attention focused on 35 elderly patients in St. Rita's Nursing Home who did not survive the sudden surge of floodwaters.

As Katrina approached the east side of New Orleans, hurricane winds began to shift from east to northeast to north over the city, causing the storm surge in Lake Pontchartrain to press against the Pontchartrain levees, especially into the "drainage" canals in the city that are open to the lake. Between 6 and 10 a.m. levees and floodwalls along the Industrial Canal, the New London Canal, and the 17th Street Canal were breached or collapsed suddenly, sending walls of floodwater into adjacent neighborhoods of the city, especially the Lower Ninth Ward, Mid-City and Broadmoor, Lakeview, and New Orleans East (see fig. 2.5). Near the levee breaches homes and buildings were swept off their foundations, and the water rose so rapidly that many residents who had not evacuated before the storm were trapped and drowned. Sections of the 8-mile I-10 causeway and bridge between New Orleans and Slidell collapsed, and at nearly the same time winds tore away portions of the Superdome roof. At this time communications between city and state agency personnel in New Orleans and the outside world failed completely, and most federal and state agencies in Washington, D.C., and Baton Rouge were unaware of the evolving catastrophe for much of the day as Katrina swept by to the east of the city.

Monday, August 29, 10 a.m. CDT: Category 3 Katrina moves rapidly northward just to the east of New Orleans and approaches the Mississippi Gulf Coast

Advisory 27 from the NHC placed Katrina 25 miles east-northeast of New Orleans and 45 miles west-southwest of Biloxi, moving north at 16 miles per hour. The central pressure had risen to 27.37 inches (927 millibars), and sustained winds had decreased to 125 miles per hour.

Even as damaging winds abated in New Orleans, hurricane-force winds continued to lash the Mississippi Gulf Coast, and a national record-high storm surge of 28 feet destroyed almost every building within the first three or more blocks of the beach from the Louisiana border eastward more than 60 miles, almost to Pascagoula. From the ground and air it looked as if a colossal class (F)-five tornado had leveled everything, even many of the centuries-old live oaks that made this coast so distinctive and charming (see fig. 2.6). The three nineteenth-century beach towns just east of New Orleans—Waveland, Bay St. Louis, and Pass Christian, all heavily damaged by Hurricane Camille in 1969—were virtually swept

Fig 2.5. Location of levee failures in New Orleans.

off the map by the storm surge. Many of the survivors owed their lives to being able to cling to tree branches just above the rising storm surge (see fig. 2.7). The surge swept through the first floor of the Hancock Medical Center in Bay St. Louis even though the center was located 2 miles inland from the beach and at an almost lofty 27 feet above sea level. Bay St. Louis was so isolated after Katrina because of destroyed bridges that it was reported that no federal first-responders reached the town until Friday, September 2.

Fig 2.6. Oblique air photo of the Biloxi, Mississippi, coast before (September 1998) and after (31 August 2005) Hurricane Katrina.

Fig 2.7. Oblique air photo of the Waveland, Mississippi, coast before (September 1998) and after (31 August 2005) Hurricane Katrina.

Monday, August 29, 10 p.m. CDT: TS Katrina weakens rapidly as it moves northward over Mississippi

Advisory 29 located TS Katrina over Columbus, Mississippi, moving north-northeast at 22 miles per hour, with sustained winds of 60 miles per hour. The central pressure had risen to 28.73 inches (973 millibars), with rapidly clearing skies, light winds, and typical sultry conditions over New Orleans and the Mississippi Gulf Coast. Because of the almost complete failure of communication networks, those outside the affected areas, particularly state and federal administrations, knew little about the loss of lives and property in New Orleans and along the Mississippi Gulf Coast, or about the life-threatening situations of immediate survivors. Along the Mississippi Coast, the return flow of the initial surge of seawater back to the Gulf was so strong that additional damage to roads and buildings occurred in many locations.

Tuesday, August 30: The massive extent of the disaster begins to be revealed

In New Orleans, media personnel stranded in hotels in the French Quarter or flown in quickly from across the nation via small airplanes, helicopters, and boats began to provide first-hand accounts of the extent of the flooding, massive destruction of buildings, and dire straits of survivors on rooftops and elevated highway structures. At seven hospitals alone, more than 7,000 patients, staff, and neighbors were in need of rescue by helicopters or boats. As water from the breeched canals continued to rise, residents focused on reaching the Superdome, the designated shelter of last resort, or the Convention Center, into which about 20,000 people had stormed even though it was completely unprepared to serve as an emergency shelter. In contrast to the Mississippi Gulf Coast, where most of the storm surge drained back into waterways and the Gulf, New Orleans floodwaters from the breached canal levees continued to rise throughout the day, further complicating rescue efforts and allowing looting to continue in many areas without police or National Guard protection. In New Orleans most of the initial search and rescue operations were carried out by up to 4,000 personnel from the U.S. Coast Guard, the Louisiana Department of Wildlife and Fisheries, and volunteers, the so-called "Cajun Navy," with their own boats. With most of the media attention focused on New Orleans, the survivors along much of the Mississippi Coast were largely isolated and left to cope on their own. At Bay St. Louis, for example, the first federal disaster responders did not reach the town until Friday morning, more than three days after the initial onslaught of destruction.

Wednesday, August 31: Floodwater levels stabilize in New Orleans

It was not possible to close the breeched levees, but floodwaters in the city finally came into equilibrium with Lake Pontchartrain; about 80 percent

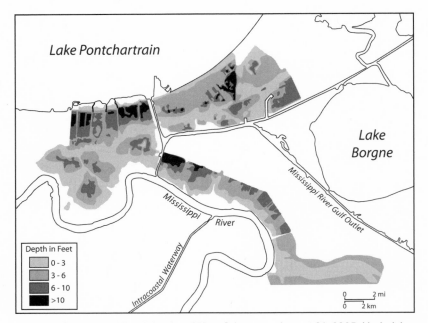

Fig. 2.8. Generalized flood depths of New Orleans on August 31, 2005. Underlying data are from NOAA at http://www.katrina.noaa.gov/maps/maps.html.

of the city was flooded up to maximum depths of more than 10 feet (see fig. 2.8), perhaps even reaching near 15- to 20-foot water depths in parts of the Lower Ninth Ward. With few flood victims able to evacuate New Orleans, the number of refugees at the Superdome increased to about 26,000, resulting in a "lock-out." New arrivals were forced to remain on nearby elevated highways in the hot sun without bathroom facilities or organized supplies of food or water, or to move on to the Convention Center, where conditions were even worse. TV pictures of people stranded outside of the Superdome and Convention Center were broadcast nationwide and around the world, presenting viewers with third-world-like images of disaster survivors without organized federal and state support, and these were followed by even more images of the aftermath (see figs. 2.9 and 2.10). The negative image of Louisiana was amplified when television stations showed Gretna police with guns blocking refugees who were trying to flee from flooded New Orleans to largely dry Gretna by crossing a bridge over the Mississippi River.

Fig 2.9. Roof surface peeled off of Superdome, New Orleans.

Thursday, September 1: Organized evacuations finally begin

Floodwater levels in New Orleans finally began to recede slightly back through the breeched canal levees as the high water levels in Lake Pontchartrain drained back out to the Gulf. Organized evacuations were initiated, and by the end of the day more than 10,000 evacuees had departed for destinations such as the Astrodome in Houston, Texas. Nevertheless, thousands of displaced persons continued to be stranded in New Orleans without food and water in the sweltering heat and humidity of late summer. An estimated 4,000 were on the raised sections of I-10 in the vicinity of the still-full Superdome, and more than 15,000 remained in the Convention Center.

Friday, September 2: President Bush briefly visits the Gulf Coast, and the pace of evacuations increases

President Bush briefly visited Mobile, Biloxi, and New Orleans, and promised federal aid and support for evacuees and the restoration of the destroyed coastal towns and cities. From New Orleans, evacuations by

Fig 2.10. Flooding mosaic: 17th Street Canal breach (*top left*); car that floated left behind in Mereaux, Louisiana (*top right*); bridge destroyed over Bay St. Louis, Mississippi (*bottom left*); flooding at the childhood home of Barry Keim in Chalmette, Louisiana (*bottom right*).

buses and trains began to impact the numbers stranded in and around the Superdome, but conditions at the Convention Center were largely unchanged until as many as 1,000 National Guard troops began to take control of the enormous convention facility. Most of the thousands remaining there had to wait one more day before evacuation.

On Saturday, September 3, the last evacuees left the Superdome and Convention Center. National Guard, Regular Army, and Marine troops were able to take control of New Orleans and bring a measure of order to the city and the Gulf Coast. It has been estimated that in the previous three days about 41,000 people were evacuated directly from the Superdome and its immediate surroundings, and an additional 20,000 from the Convention Center. Close to 250,000 evacuees were in officially designated shelters and hotels in Texas, and large numbers also overfilled facilities in Baton Rouge, Jackson, Birmingham, and Atlanta. In fact, the diaspora of evacuees was so extensive that New Orleans refugees were found in almost all fifty states. Some families were inadvertently separated for days, if not weeks.

Altogether in the three states of Louisiana, Mississippi, and Alabama, more than 1.7 million people had been displaced from their homes. On August 24, 2006, the *New York Times* reported that on the basis of change of address forms with the U.S. Post Office, of household units in the New Orleans metropolitan region prior to Katrina, 52,000 were located elsewhere in southern Louisiana, especially Baton Rouge; 34,000 in Houston; 15,000 in Dallas-Fort Worth; and 11,000 in Atlanta. Two years later it was estimated that not quite half of the pre-Katrina population had returned to New Orleans, with even fewer returning to Plaquemines and St. Bernard Parishes.

By Monday, September 5, the levee breaches were closed, and the task of pumping New Orleans dry began. About 80 percent of the city was flooded. To add to the catastrophe, category four Hurricane Rita passed about two hundred miles southwest of New Orleans on Friday, September 23, on its way to landfall in Cameron Parish, Louisiana, just a few miles east of the Texas border.

Unfortunately, the seven-foot storm surge in New Orleans was high enough to break through some of the partially-restored levees along the Industrial Canal, and floodwaters again poured into the Lower Ninth Ward, Arabi, and Chalmette on the east side of the canal, and to a lesser degree into Gentilly on the west side. The two photos in figure 2.11 were taken in the Lower Ninth Ward of New Orleans shortly after Hurricane Katrina (top) and after Hurricane Rita (bottom). The top image shows the infamous barge that was sucked through the colossal levee breach along the Industrial Canal, with a small bus located tens of feet away. In the bottom figure, note how the barge was obviously floating in the Rita-induced surge and eventually grounded on top of the small bus. It was not until October 11 that the Corps of Engineers declared that New Orleans was officially dry, forty-three days after the initial flooding on August 29. Fig. 2.12 shows several scenes of damage in the devastated Lower Ninth Ward region.

Fig. 2.11. The top photograph was taken shortly after Hurricane Katrina; the bottom photo was taken shortly after Hurricane Rita. Note the juxtaposition of the barge and bus in the two photos, clearly showing that surge from Rita reflooded the Lower Ninth Ward of New Orleans causing the barge to relocate.

Fig. 2.12. Scenes from around the Ninth Ward of New Orleans following Hurricane Katrina.

Although much is known about the specific timeline of events in New Orleans, less is known about what transpired in many of the suburbs in Plaquemines, St. Bernard, and St. Tammany Parishes, all of which were hard hit by the storm. By August 2006, only about 6,000 of the pre-Katrina population of about 19,000 residents of the lower, or downstream, sections of Plaquemines Parish had returned. Despite the grim conditions, many planned to return, and this recovery will take decades (see fig. 2.13).

The Mississippi Coast also bore the brunt of the storm, falling in the most powerful quadrant of the hurricane. This brought even stronger winds and higher surge to Mississippi than those experienced in New Orleans. There was a 20-mile swath of surge between 24 and 28 feet that hit the western Mississippi coastline, centered on Bay St. Louis, and a 17- to

Fig. 2.13. Defiant evacuees from St. Bernard Parish vow to return after the ravages of Hurricane Katrina.

22-foot surge between Gulfport and Pascagoula. The surge penetrated 6 miles inland along much of the coast, and up to 10–12 miles along rivers. Most of the homes within several blocks of the Mississippi Coast were swept off their foundations. Furthermore, all of the twelve floating gambling casinos on the Mississippi Gulf Coast, with a thirteenth about to open in early September 2005, were heavily damaged or destroyed. But the Mississippi legislature soon thereafter voted to allow the casinos to be rebuilt on land, and three were back in operation by the end of the year. By the end of 2007, eleven of the thirteen casinos were rebuilt and operating.

The following list summarizes damage in Mississippi:

- 90 percent of structures within half a mile of the coast destroyed
- 60 percent of housing in the three coastal counties (Hancock, Harrison, and Jackson) damaged or destroyed
- 90 percent of Pascagoula, seventy-five miles east of landfall, flooded by surge
- 27- to 28-foot storm surge recorded at Bay St. Louis—the highest in U.S. history
- Surge penetrated six miles inland and up to twelve miles along bays and rivers
- Two major bridges on coastal highway US 90 destroyed, Bay St. Louis-Pass Christian and Biloxi-Ocean Springs, resulting in long detours
- Gulf-front casinos and hotels heavily damaged, some of the casino barges swept inland across Highway 90; casinos have returned strongly, but thousands of units of affordable housing for working class people and families have not
- Limited labor, personnel, and funding for reconstruction
- Oil platform ripped from foundation and swept to beach on Dauphin Island in Alabama
- Port facilities at Gulfport destroyed and hundreds of eighteen-wheeler trailers swept inland

An absolutely exact count of the loss of human lives associated with Katrina is not possible, especially because cause of death was often medical rather than drowning. Nevertheless, summary counts six months later amounted to over 1,100 in Louisiana, mostly in and around New Orleans, at least 225 in Mississippi, and 16 in Florida, with a national total of about 1,350. A considerable number of the drownings in New Orleans have to be attributed to the lack of a timely coordinated federal, state, and local first-responder program that should have been activated within hours after the departure of Katrina Monday afternoon. By comparison with the Galveston Hurricane 105 years earlier, technology enabled forecasters in 2005 to track Katrina precisely in terms of location and intensity, but predictions beyond two to three days were uncertain enough to encourage some residents to take the chance that Katrina would go elsewhere or weaken. From Pensacola, Florida, westward to beyond New Orleans, it has been estimated that more than 250,000 homes, condominiums, and apartments were heavily damaged or destroyed, and property and infrastructure losses have been estimated to be as much as $75 billion, making Katrina by far the most costly natural disaster in American history.

The Galveston Hurricane of 1900 and Hurricane Katrina in 2005 were the two most catastrophic hurricanes along the shores of the Gulf of Mexico in the past 108 years. How often can Gulf coastal residents expect visitations by tropical storms and hurricanes? Looking to the past to anticipate the future, we provide some answers to this important question based on more than one hundred years of the geography and history of tropical storm and hurricane strikes at coastal places counter-clockwise from Key West to the Yucatan in Mexico.

3

COUNTING TROPICAL STORMS AND HURRICANES
OVER THE GULF OF MEXICO

In the 105 years separating the Galveston Hurricane and Hurricane Katrina, many other tropical storms and hurricanes came on shore around the Gulf, a few barely newsworthy and others almost as destructive of life and property as the 1900 and 2005 monster hurricanes. Furthermore, there are historical records of other devastating hurricanes along the northern Gulf Coast prior to 1900. Two of these storms completely destroyed Indianola, Texas, a small but prosperous seaport 14 miles inland from the open Gulf on Matagorda Bay and about 120 miles southwest of Galveston. The first hurricane struck on September 16, 1875, with three-quarters of the buildings destroyed and 176 inhabitants drowned. The second hurricane hit eleven years later on August 20, 1886, and was so destructive that Indianola was never rebuilt. A little more than seven years later, on October 2, 1893, an unanticipated hurricane inundated the barrier island of Cheniere Caminada, southwest of New Orleans. About half of the town's population drowned, and altogether about two thousand people died, mostly in southeastern Louisiana (see chapter 5).

We explore now the number of tropical storm and hurricane strikes around the shores of the Gulf of Mexico over the past one hundred-plus years. Each May, before the upcoming storm season begins on June 1, several federal and private-sector hurricane groups issue their seasonal predictions of the number of Atlantic Basin storms including the number of tropical storms, hurricanes, and severe hurricanes (categories three through five) (see table 2.1). Although the forecasters give some attention to broad geographical regions—the Atlantic Coast, Caribbean, or Gulf, for example—the media tend to focus on storm counts for the entire Atlantic Basin. Therefore we turn first to seasonal counts of storms for the entire Atlantic Basin and then to the number of these storms that have been observed over the waters of the Gulf.

National Hurricane Center (NHC) personnel have identified 1,112 tropical storms and hurricanes over the entire North Atlantic Basin, including the Caribbean Sea and the Gulf of Mexico, between 1886 and 2005, with a long-term seasonal average of 9.3 storms. Figure 3.1 shows the average number of storms by five-year periods, ranging from a maximum of seventeen storms per season between 2001 and 2005 down to an average of only five storms per season between 1911 and 1915. Quite remarkable is the longer-term average of only 5.5 storms per season between 1911 and 1930, the relatively steady average counts of 9.7 storms per season between 1936 and 1995, and the very striking increase beginning with 1996.

It has been recognized, however, that the count of tropical storms and hurricanes over the Atlantic Basin is incomplete, especially for the years prior to the later 1930s. Storms that remained over the open Atlantic were recognized only when ships were caught inadvertently near or within the storms. In the late 1930s, increasing trans-Atlantic air and sea traffic, especially because of World War II, resulted in more storm identifications, and with the introduction of routine satellite observations in the 1960s, there has been minimal possibility of storms remaining unidentified. Therefore, Christopher Landsea at the NHC has recommended upward statistical adjustments of 3.2 storms per season to the Atlantic Basin storm records prior to 1966 to take into account the unidentified open-ocean storms. Landsea's recommended adjustment increases the number of estimated storms during the less active period between 1911 and 1930 to about nine per season, but the adjustment also increases the annual number of storms before and after this period equally, leaving 1911 to 1930 a period with significantly less activity than the long-term average. Nevertheless, we have not adjusted the storm counts in figure 3.1 to take into account Landsea's estimates of these undetected mid-ocean storms, and the figure represents the official NHC count of identified tropical storms and hurricanes.

Figure 3.1 also shows the number of these same storms with tracks over the Gulf of Mexico. The 120-year total adds up to 395 storms, or 36 percent of all of the Atlantic storms, averaging 3.3 storms per year. Two five-year periods, 1931–1935 and 2001–2005, tied for most frequent storms, averaging six per season, with more frequent activity between 1886 and 1910, between 1931 and 1970, and again between 2001 and 2005. Between 1926 and 1930 there were only 1.4 storms per season on

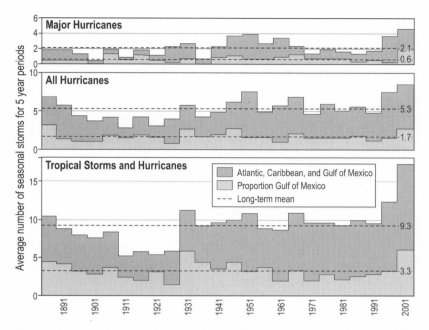

Fig. 3.1. Tropical storm, hurricane, and severe hurricane frequencies averaged for a season by 5-year (pentad) periods from 1886 through 2005. The entire Atlantic Basin is represented with the proportional contribution of Gulf storms.

average. Because of the geographical boundaries of the Gulf, very few storms over the Gulf during the early years of the twentieth century must have gone undetected, and our count of storms over the Gulf based on the official storm tracks of the NHC has to be reasonably representative of activity beginning with 1886.

For the entire Atlantic Basin, 639 storms, or a little more than half of all storms, attained hurricane status, a seasonal average of 5.3 hurricanes for the 120 years. Figure 3.1 shows more frequent hurricanes between 1886 and 1895, between 1946 and 1965, and during the most recent ten years beginning with 1996, when the seasonal average reached 8.1. Between 1896 and 1925 hurricane activity was much less frequent, with a seasonal average of only 3.7 hurricanes, again partially affected by Landsea's evaluation of undercounting during the first half of the twentieth century.

Figure 3.1 also shows that 201 hurricanes, about one-third of all the Atlantic hurricanes, had tracks over the Gulf of Mexico; hence 51 percent of all storms over the Gulf reached hurricane status with an average of 1.7 hurricanes per season. Six of the five-year periods averaged at least two hurricanes per season: 1886–1890, 16 hurricanes; 1946–1950, 14 hurricanes; 1931–1935 and 2001–2005, 13 hurricanes; and 1941–1945 and 1966–1970, 10 hurricanes. There were only four hurricanes between 1926 and 1930.

For the entire Atlantic Basin, figure 3.1 also shows that 248 storms, 22 percent of all storms, reached major hurricane status of category three or greater, with a seasonal average of 2.1 events. Two extended active periods stand out, 1926–1970, with the exception of 1936–1940, and 1996–2005. About one quarter of the major Atlantic hurricanes, 72 storms, had tracks over the Gulf of Mexico. From a different perspective, 72 of the 395 tropical storms and hurricanes over the Gulf reached major hurricane status. There were nine between 2001 and 2005, six between 1906 and 1910 and again between 1966 and 1970, and five between 1946 and 1950.

The first category five hurricane over the Atlantic was not identified until 1928, with an average over 120 years of only one category five hurricane every four years. For the last sixty years beginning with 1946, however, the frequency of category five storms has increased to nearly one storm every two years on the average, with six in the five years between 2001 and 2005!

Of the 30 category five hurricanes over the Atlantic Basin beginning in 1928, eleven occurred over the Gulf of Mexico. These are the Labor Day Hurricane in the Florida Keys in 1935, Ethel in 1960, Carla in 1961, Beulah in 1967, Camille in 1969, Anita in 1977, Allen in 1980, Ivan in 2004, and Katrina, Rita, and Wilma in 2005. These data suggest that the second half of the twentieth century had more frequent major hurricanes than the first half, even though limited observational technologies have certainly affected the intensity estimates of storms in the decades before routine introductions of radars, satellites, hurricane-hunter flights into storm circulations, and reevaluations of historical storm data by personnel of the NHC.

We turn now to the historical climatology of tropical storm and hurricane strikes at specific coastal locations beginning in 1901. In figure 3.2 we have evaluated tropical storm and hurricane strikes at 57 coastal sites, 30 around the shores of the Gulf from Key West to the Yucatan and an additional 27 sites along the Atlantic Coast from Key Largo to Eastport, Maine, using our model of the average extent of storm winds in terms of

Saffir-Simpson intensity categories. The storm track data are for every six hours as reported by the NHC, including precise historical locations of storm centers and estimates of maximum wind speeds; these data sets are available at several sites on the Internet.

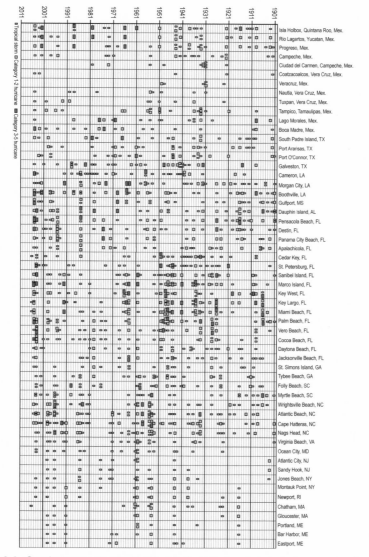

Fig. 3.2. Spatiotemporal pattern of tropical storm, hurricane, and severe hurricane strikes along the Gulf and Atlantic coastlines from 1901 through 2007.

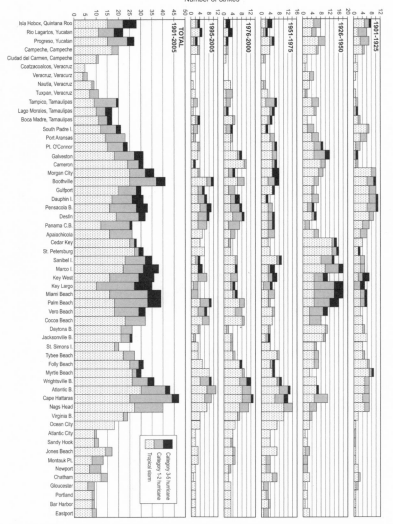

Fig. 3.3. Spatiotemporal pattern of tropical storm, hurricane, and severe hurricane strikes along the Gulf and Atlantic coastlines broken down into quarter centuries (1901–1925, 1926–1950, etc.), along with the period 1995–2005.

Figure 3.3 shows the strike data at 57 sites around the Gulf, from the Yucatan to Key West, and then along the Atlantic Coast northward to Maine for comparative purposes. The data are presented by Saffir-Simpson categories and organized into four twenty-five-year periods, 1901–

1925, 1926–1950, 1951–1975, 1976–2000, and the very active eleven years from 1995 to 2005. The entire 105-year record is summarized at the bottom of figure 3.3. Comparisons can be seen by looking at geographical locations horizontally across the chart or through time by scanning vertically downward for individual places or regional clusters along the coasts. There are much smaller chances of strike bias at landfall in these years because, even at the beginning of the twentieth century, extended stretches of coastline without any permanent settlements were less common than areas of the open Atlantic without ships during a hurricane event.

Look first at the striking geographical pattern of tropical storm and hurricane strikes for 105 years along the shores of the Gulf of Mexico at the bottom of figure 3.3. There have been 25 or more tropical storm and hurricane strikes in 105 years at each city from Key West north and west to Galveston, Texas, and at two of the four coastal cities in the Yucatan. The highest number of strikes, 41, is at Boothville, Louisiana, southeast of New Orleans near the mouth of the Mississippi River. Thirty-five or more hit Morgan City, Louisiana, and southwestern Florida from Sanibel Island south to Key West. Along the Texas coast south of Galveston and the Mexican coast there have been fewer strikes, with no more than 11 strikes between Tuxpan and Ciudad del Carmen; at Coatzacoalcos on the southern coast of the Bay of Campeche, there have been only five tropical storm strikes, and no hurricanes, in 105 years!

Along the Atlantic Coast by comparison, tropical storm and hurricane strikes have been most frequent in southern Florida and along the Outer Banks of North Carolina. The maximum strike count from Maine to the Yucatan has occurred at Cape Hatteras with 47 strikes, almost one every other season on average. Southern Florida is especially exposed to hurricanes sweeping westward from the Atlantic and northwestward from the Caribbean, and the Outer Banks to storms curving to the north and northeast along the Atlantic coast. The coastal configuration from Daytona Beach to Myrtle Beach shelters these shores from many direct strikes, with only 20 storm strikes at St. Simons Island in Georgia. Northward from the Outer Banks the number of strikes decreases dramatically with only nine to 15 strikes at all sites from Atlantic City northward because cooler water temperatures and frequent upper-air steering move storms farther offshore. The exception is Jones Beach on Long Island with 17 strikes.

The geographical pattern of hurricane strikes can also be visualized in figure 3.3. For the areas along the Gulf Coast, at least five to ten hurricane strikes over 105 years is the rule, roughly one every ten to twenty years on average. The most frequent strikes were at Key West, 20, and Marco Island, 16. More than ten struck each of the areas from Morgan City eastward to Panama City, with Pensacola Beach recording the most at 17. Galveston and Isla Holbox on the Yucatan both experienced 13 strikes. The low frequency exceptions are the Mexican coast from the Yucatan westward around to Tuxpan, with two or less between Ciudad del Carman and Veracruz and none at Coatzacoalcos, and in Florida from east of Apalachicola southward to St. Petersburg, with less than five strikes.

Along the Atlantic Coast, by comparison, there have been 20 or more hurricane strikes at Cape Hatteras and from Key West northward to Palm Beach, with Key Largo receiving the maximum of 25 hurricane strikes, about once every four years on average. There have been less than ten strikes from Cocoa Beach north to Myrtle Beach, with only two at St. Simons Island. There have been few strikes north of Nags Head, with none for the 105 years at Ocean City, Maryland.

Major hurricanes, categories three through five, along the Gulf Coast, have recurred most frequently in southwestern Florida, with seven at Key West and Marco Island; along the central Gulf Coast, with five at Pensacola Beach and Dauphin Island; and six at Isla Holbox on the northern coast of the Yucatan. Most other sites have one to four strikes, but our analysis shows no major hurricane strikes along the shores of the Bay of Campeche from Campeche around to Tuxpan.

Along the Atlantic Coast, by comparison, major hurricane strikes have occurred most frequently from Key West northward to Palm Beach, with six or more strikes at each site and a maximum of eight at Key Largo. A higher proportion of hurricane strikes in southern Florida are classified as major than anywhere else from the Yucatan to Maine. There have been no major hurricane strikes from Cocoa Beach northward to Tybee Beach (Savannah), Georgia, except for one at Jacksonville Beach. From Folly Beach (Charleston), South Carolina, northward to Cape Hatteras there have been two or three strikes at each site, but no major hurricane strikes from Nags Head, North Carolina, to Maine in the 105-year record, though the Hurricane of 1938 may be an exception.

Figure 3.3 also shows the frequencies of strikes for twenty-five-year periods, 1901–1925, 1926–1950, 1951–1975, and 1976–2000, as well as the recent eleven-year period from 1995 through 2005, which was the most

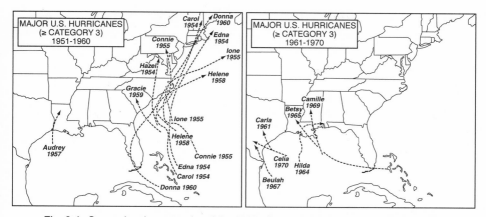

Fig. 3.4. Severe hurricane tracks of the 1950s (focused on the U.S. East Coast) and 1960s (centered on the Gulf Coast).

active strike run in the historical record for many of the sites around the Gulf of Mexico. The arbitrary selection of twenty-five-year periods is employed here to emphasize the varying geographical strike patterns through the years.

Figure 3.3 shows the frequent reoccurrence of tropical storms and hurricanes in 1901–1925 across the eastern half of the northern Gulf from Morgan City to Apalachicola and the Yucatan. At the same, there were relatively infrequent strikes along the Bay of Campeche northward to Texas and western Louisiana. The twenty-five-year period between 1926 and 1950 was very active along the western and eastern Gulf coasts, particularly from Cedar Key to Key West, and along the northwestern coast centered on Galveston.

The 1951–1975 period was moderately active almost everywhere, with several major hurricane strikes in the Yucatan, Louisiana, and southwest Florida, but was relatively inactive along the shores of the Bay of Campeche. The patterns of ocean water temperatures and atmospheric pressure over the open Atlantic strongly affect storm track patterns over the Atlantic, Caribbean, and Gulf. Figure 3.4 illustrates how most major hurricane strikes for the United States between 1951 and 1960 occurred along the East Coast, with only one strike over the Gulf, but, by contrast, all of the major hurricane strikes during 1961–1970 occurred along the shores of the Gulf.

For 1976–2000 there were fewer major hurricane strikes overall, but more frequent tropical storm and hurricane strikes along the northern Gulf Coast and less frequent activity over the western Gulf and the coasts of Mexico and Texas. Finally, the strike focus for the eleven-year period between 1995 and 2005 was on the northern Gulf Coast from eastern Texas to the Florida Panhandle, with numerous tropical storms and several notorious major hurricanes in 2004 and 2005—Ivan, Dennis, Katrina, and Rita. During the same two seasons two other major hurricanes, Charlie and Wilma, ravaged much of the southwestern Florida coast south of St. Petersburg.

The individual tropical storm and hurricane strike data around the Gulf of Mexico are displayed in figure 3.5. Every strike for 107 years between 1901 and 2007 at thirty coastal sites from the Yucatan to Key West is included in this chart. The strikes have been divided into tropical storms, category one and two hurricanes, and major hurricanes (categories three through five), with larger symbols for the increasing severity of the strikes by category. Altogether at the thirty coastal sites around the Gulf from 1901 through 2007, there have been 468 tropical storm strikes, 205 hurricane category one and two strikes, and 73 major hurricane strikes, for a grand total of 746 strikes.

Figure 3.5 provides a comprehensive visual regional perception of the active and inactive runs of years in terms of storm strikes from the Yucatan around the shores of the Gulf to Key West for more than one hundred years, 1901–2007. The figure also provides specific records of strike patterns over time at each site. Around the Gulf, hurricanes come in bursts, with more in some years and fewer in others. For example, strikes tended to concentrate on the eastern half of the northern Gulf Coast from 1901 to 1926, with infrequent strikes along the western half of the northern Gulf Coast, Texas, and Mexico except for the Yucatan. Another example is the very frequent strikes along the western coast of Florida between 1925 and 1950, the Texas coast between 1932 and 1949, and the Yucatan coasts between 1932 and 1944. A final example is the clustering of major hurricanes along the northern Gulf Coast and southwestern Florida during the 1960s and again in 1995–2005.

There is so much detail in figure 3.5 that the reader would do best to focus on individual places of personal interest. Nevertheless, several interesting examples include the cluster of ten strikes in twelve years at Galveston between 1938 and 1949 and the eleven years at Pensacola Beach between 1995 and 2005, which included four tropical storms, three cat-

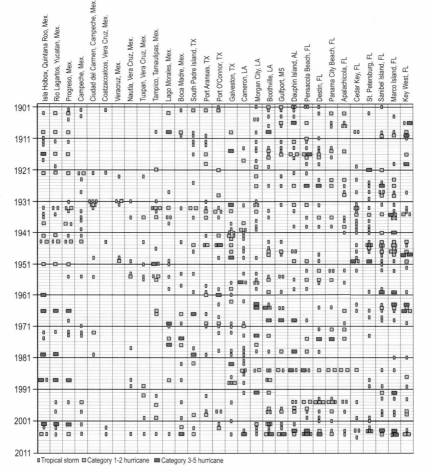

Fig. 3.5. Spatiotemporal pattern of tropical storm, hurricane, and severe hurricane strikes along the Gulf Coast, 1901–2005.

egory one or two hurricanes, and two major hurricanes. At Ciudad del Carmen on the Bay of Campeche, there were six tropical storm strikes in three years between 1931 and 1933, and only four other tropical storms and one hurricane during the entire 106-year record. At Veracruz there has not been a single tropical storm or hurricane strike since 1950. And, finally, at the three northern sites on the Yucatan there was only one major hurricane strike between 1901 and 1960, but twelve strikes from 1961 to 2007.

| 4 |

Hurricane Basics

Tropical storms and hurricanes receive considerable media attention because of the extraordinary damage they cause. Some examples include the heavy rainfall produced by Tropical Storm Allison over Houston in 2001, the catastrophic wind damage from Hurricane Andrew in South Florida, and the incredible storm surge induced by Hurricane Katrina in Louisiana and Mississippi. Clearly, these systems represent the most dangerous meteorological hazards over the oceans, inflicting destruction within the coastal zone to both natural landscapes and the built environment.

We now consider Gulf of Mexico hurricanes from the perspectives of where and why these storms form: their seasonality, their regions of development, prevalent storm tracks, why some years and decades are more active than others, and how predictions are made for upcoming hurricane seasons. Note that North Atlantic hurricanes comprise only 11 percent of global hurricanes and that Gulf hurricanes make up only 31 percent of North Atlantic hurricanes.

In the Gulf, we call storms "hurricanes"—from *huracan*, the word for a Carib Indian god of evil. Other regions use different names for the same meteorological feature. They are called "typhoons" in the western North Pacific and "tropical cyclones" in the South Pacific and Indian Oceans. Figure 4.1 shows the breeding grounds for these storms across the globe.

Hurricane Formation

Tropical storms and hurricanes form to help redistribute energy across the earth's surface, oceans, and atmosphere. They form over tropical and subtropical ocean surfaces where heat accumulates during summer and early autumn. In the Gulf and Atlantic, the energy is then redistributed

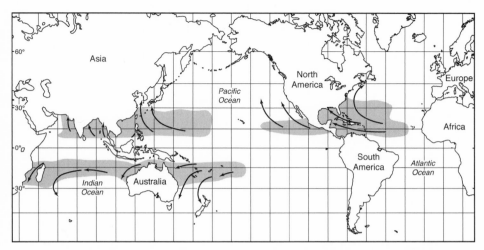

Fig. 4.1. Breeding grounds (in gray) for tropical storms and hurricanes around the world.

northward by atmospheric and oceanic currents in an effort to maintain a balance of heat and energy across these latitudes.

Hurricane formation requires a suite of oceanographic and meteorological conditions. First, sea surface temperatures (SSTs) greater than or equal to 80° F are needed to provide energy for the storms to develop and intensify, and warmer ocean surfaces lead to greater potential for hurricane development. Second, evaporation rates off the ocean must be high. The evaporation process transfers energy from the ocean to the atmosphere, where it is stored as latent energy and then later released as sensible (or measureable) heat when condensation occurs in the form of clouds. This release of latent (stored) energy to sensible (heat) energy makes storms even more powerful by creating a very unstable atmospheric environment. Unstable in this context means that rapid uplift of air and moisture within the atmosphere is possible.

Third, upper air flow at approximately 25,000–50,000 feet must allow the rising moist tropical air around the center of the storm to vent aloft outward from the center. Otherwise, strong winds at high levels of the atmosphere can create wind shear, and storms are not allowed to vent aloft. Wind shear can be defined as an abrupt change in wind speed or direction over a short distance, and it can be either vertical or horizontal. In this case, horizontal wind shear serves to tear apart the hurricane by limiting, or preventing, ascent of air needed to support the storm dynamics.

Finally, storms in the northern hemisphere generally form between 5° N and 25° N latitude. The region between the equator and 5° N latitude has weak to nonexistent Coriolis forcing, thus preventing the necessary rotation needed to produce the cyclone. Coriolis forcing causes a deflection in the direction of moving air, thus aiding in formation of cyclones (and anticyclones). When the factors noted here are in place, potential is high for storm formation and further intensification.

The formation of hurricanes typically begins as a cluster of thunderstorms. Although some Gulf storms are bred over Gulf waters, many storms can be tracked to either the Caribbean or Atlantic, and occasionally the African continent, such as in the Sahel in Africa. Once a cluster of thunderstorms is over water, they typically form an area of weak low pressure causing a perturbation in the pressure pattern—called a tropical wave (see fig. 4.2). These waves produce convergence of surface air currents on the east side of the wave axis and divergence on the west side, thereby enhancing uplift in the atmosphere where convergence takes place. Furthermore, the wave serves as the impetus for rotation, and eventually a closed circulation pattern as the storm intensifies. In a typical year, the North Atlantic Basin may have over one hundred tropical waves form, any one of which may affect the Gulf of Mexico. Only a small percentage of these waves will develop into a tropical storm, and even fewer into a hurricane.

After a closed circulation forms, the storm is called a tropical depression, with a central location anchoring the circulation. Once wind speeds around this low pressure reach 38 miles per hour, the storm becomes a tropical storm. When wind speeds reach 74 miles per hour, it becomes a hurricane with a calm "eye" in the center and an eye wall around the eye. Here the highest winds and rainfall amounts are found. Hurricanes also consist of convective feeder bands rotating around the eye, with weak downdrafts in between the feeder bands (see fig. 4.3). The eye is dominated by downdrafts of air and is generally cloud-free, with relatively calm winds at the surface. For example, figure 4.4 shows the eye of Hurricane Andrew on August 25, 1992, with the eye clearly visible only hours before it made landfall near Morgan City, Louisiana. In some images,

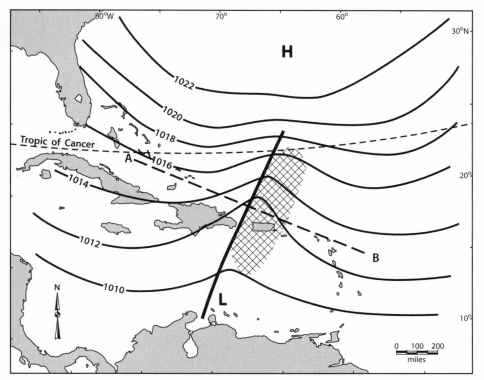

Fig. 4.2. Example of a tropical wave located east of the Gulf of Mexico.

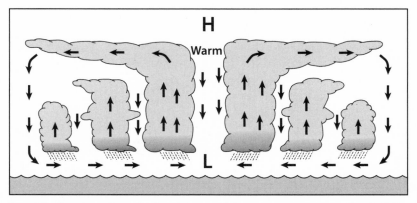

Fig. 4.3. Cross-section of a hurricane, with low pressure and the "eye" at the storm's center.

Fig. 4.4. Hurricane Andrew over the Gulf of Mexico on 25 August 1992.

the ocean surface is even visible within the hurricane's eye. After reaching hurricane strength, storms are then measured on the Saffir-Simpson scale, ranging from category one (weak) to category five (with winds over 155 miles per hour) (see table 2.1).

Hurricane Seasonality

The seasonality and intensity of tropical storms and hurricanes are primarily driven by SSTs. In the Gulf, hurricane season officially spans the six-month window from June 1 to November 30, though storms do occasionally occur outside of this window. In fact, in the Gulf Basin, fifteen storms have formed in May from 1886 through 2006 (see fig. 4.5), and there have been others from January through April. The season also does not suddenly come to a halt on the last day in November—evidenced by nine storms that have formed in the Atlantic Basin in December over the same period, though none ventured into the Gulf.

Frequencies and intensities of storms generally follow the pattern of SSTs, both of which increase from May through September, peaking near September 10, before beginning to taper off (see fig. 4.6). The core of the season ranges from mid-August through the first week of October. The highest frequencies of storms occur during this period, including some of the most famous Gulf hurricanes—the Galveston Hurricane of 1900, Camille, Andrew, Katrina, and Rita.

Gulf of Mexico storms can develop directly in the Gulf, but most often they travel to the Gulf after forming in the Atlantic or Caribbean. During development, most of these North Atlantic storms travel westward at tropical latitudes, steered by persistent trade winds that blow from east to west. As storms drift to higher latitudes, they get passed from the trade winds over to the westerlies, where storm tracks tend to move from west to east. In essence, the storms track around the Atlantic Subtropical High, also called the Bermuda High or the Azores High. The strength and position of the Atlantic Subtropical High thus has a notable impact on hurricane tracks. When this high pressure feature is positioned farther west than normal in the Atlantic and is stronger than normal, the Gulf of Mexico can anticipate more than its usual share of storms that season. Figure 4.5 depicts all the tropical storm and hurricane tracks from 1886 through 2006 for each month from May to December. The parabolic curved path around the Atlantic Subtropical High is clearly evident as storms drift from east to west in the trade wind belt, then drift poleward until steered by the westerlies.

In May and June, the Caribbean and Gulf are much more active than the Atlantic because these waters tend to warm more quickly than the larger, tropical Atlantic Ocean. In July, as the deeper Atlantic water warms, storms begin forming there, and the migration around the Atlantic Subtropical High becomes more apparent. This path is even more noticeable in August and September, when storms can sometimes be tracked all the way from western Africa—called Cape Verde Storms because they form near the Cape Verde Islands. In October, the tracks become more erratic, as SSTs cool and the Atlantic Subtropical High weakens. In November and December, storm tracks become very irregular. A very weak

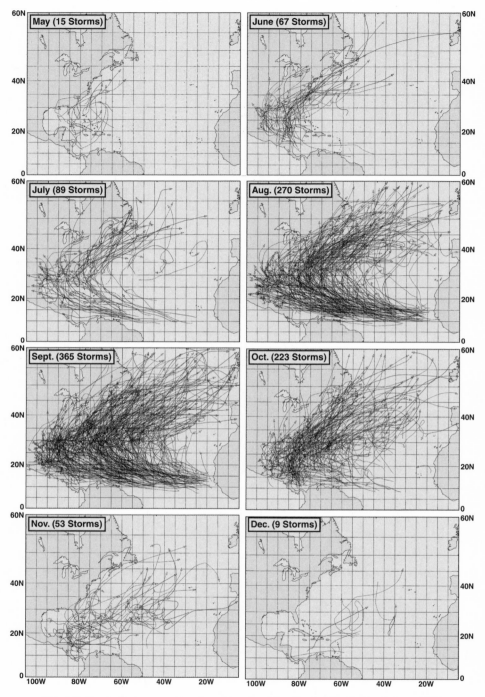

Fig. 4.5. Tropical storm and hurricane tracks of the North Atlantic Basin from 1886 through 2006 by month, May–December.

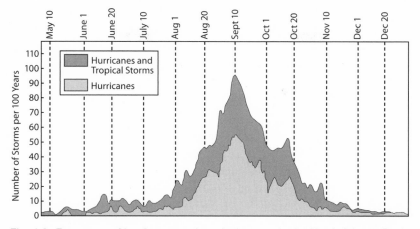

Fig. 4.6. Frequency of hurricanes and tropical storms in the North Atlantic Basin over the course of a hurricane season.

Atlantic Subtropical high, combined with traveling mid-latitude cyclones and anticyclones means that positions of the jet stream have a greater impact. Also note that there are very few storms that tracked into the Gulf in November, and there were none in December.

Gulf Loop Current

Water currents within the Gulf of Mexico can often play a role in the intensification of hurricanes within the basin. These currents serve as the mechanism to transport heat and energy and can be horizontal (across the surface) or vertical (from the surface water of the Gulf to some depth, or vice versa). One of the most important currents in the Gulf is a feature called the loop current (see fig. 4.7). The loop current is an extension of the Gulf Stream, which warms water as it travels through the Caribbean. The loop current enters the Gulf through the Yucatan Channel—between the Yucatan Peninsula and Cuba—and from there curves into the Gulf of Mexico in a clockwise direction, forming a horseshoe-shaped current, and exits the Gulf between Cuba and Florida. It brings exceptionally warm water into the Gulf, whereby hurricanes can feed off the energy. Furthermore, this warm water extends to great depths in the Gulf. This is important because hurricane winds tend to cause churning and overturning of the water over which they pass, bringing cooler water from below up to the surface—a process called upwelling. However, in this instance, the water from below is also warm despite the upwelling. Also, the loop can get truncated from the main flow of the Gulf Stream, thereby forming an eddy of warm water that can take months to dissipate as the area of relatively high SSTs propagates through the Gulf.

Two recent examples include Hurricanes Katrina and Rita in 2005. Both storms had tracks that passed over the warm waters associated with the loop current, and both experienced rapid intensification at that time. Storm development in this manner will occur only if the upper air is also conducive to the development, which was obviously the case for these two storms. Both storms entered the Gulf from the Atlantic moving nearly westward, then began to turn northward. Along Hurricane Katrina's track, the hurricane moved right over the loop current waters, where the maximum sustained winds increased to a strong category five

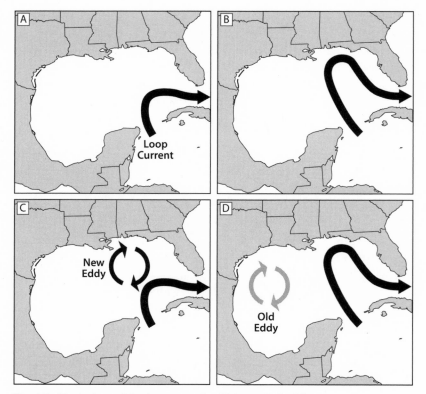

Fig. 4.7. Chronology of the loop current, with truncated eddies.

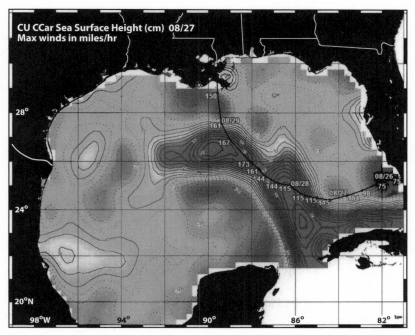

Fig. 4.8. Track of Hurricane Katrina in relation to the location of the loop current. In this case, the sea surface height is used as a surrogate for the sea surface temperature.

strength at 170–175 miles per hour (see fig. 4.8). As the hurricane moved closer to shore, it moved over cooler water, therefore decreasing in intensity before landfall. With Hurricane Rita, an eddy spun off the main loop current and was positioned just west of the northern rim of the main loop (see fig. 4.9). Rita passed over the fringe of this eddy, and some modest strengthening took place at that time.

Forecasting Hurricane Seasons

Our ability to forecast climate over months, seasons, or years has been poor, historically. However, recent research in atmospheric teleconnections has led to dramatic improvements in our understanding of how the climate system operates at these time scales. The scientific community still has a long way to go in this regard, but teleconnection research has allowed us to take a giant step forward with long-range forecasting. In es-

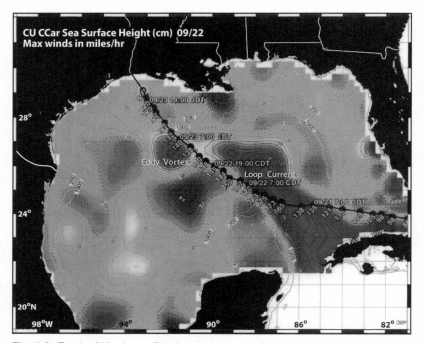

Fig. 4.9. Track of Hurricane Rita in relation to the location of the loop current. In this case, the sea surface height is used as a surrogate for the sea surface temperature.

sence, a teleconnection is an association between two or more locations, so that if something happens climatologically within one region, a predictable outcome can be expected at the teleconnected region or regions. For example, an El Niño event is the occurrence of abnormally warm sea surface temperatures in the tropical eastern Pacific Ocean. When this occurs, it affects the weather and climate in many other locations around the world—e.g., enhanced storminess in California and along the Gulf of Mexico's central coast, unusually dry conditions in northern Australia, abnormally warm winters in the northwestern United States, and decreased hurricane activity in the Atlantic Basin.

Teleconnections can also assist the scientific community in predicting hurricane seasons many months in advance. In the United States, two research units receive most of the attention for their seasonal forecasts—the Tropical Meteorology Project at Colorado State University and the National Oceanographic and Atmospheric Administration (NOAA).

The Tropical Meteorology Project was spearheaded by Dr. William "Bill" Gray, though he has given primary responsibility to one of his research associates, Philip Klotzbach. NOAA forecasts are put together by researchers from the National Hurricane Center, Climate Prediction Center, Hydrometeorological Prediction Center, and the Hurricane Research Division of the Atlantic Oceanographic and Meteorological Laboratory. With both hurricane season forecasts, an array of atmospheric teleconnection patterns are analyzed, and forecasts are then crafted from these predictors. The teleconnection predictors include the sea surface temperatures in the Atlantic Ocean—termed the Atlantic Multidecadal Oscillation (AMO)—the El Niño Southern Oscillation (ENSO) in the tropical Pacific Ocean, as well as pressure patterns and SSTs over the Tropical Atlantic and Caribbean, among other predictors.

Warm SSTs in the Atlantic Ocean are indicative of strong currents, i.e., the Gulf Stream. These strong currents transport warm water farther north than normal and, through complex cause and effect factors, also shift the west-to-east midlatitude jet stream farther north. This lessens the potential for vertical wind shear in the hurricane breeding grounds located farther south in the tropical Atlantic, Caribbean, and Gulf of Mexico. When the AMO occurs at the same time as lower pressure (and warm SSTs) over the tropical Atlantic, Caribbean, and Gulf, there is potential for very active hurricane seasons.

Another modulating factor for hurricane activity over the course of a season is ENSO. Many researchers have demonstrated that warm phase ENSO conditions—better known as El Niño—in the eastern Pacific Ocean, coupled with drought in the African Sahel, increase potential for westerly vertical wind shear over the main Atlantic breeding grounds. This creates unfavorable conditions for hurricane formation in the Atlantic; hence fewer storms will occur over the Gulf. The annual probability of a major hurricane landfall in the United States is 23 percent during a (warm phase) El Niño season, 58 percent during neutral conditions, and 63 percent during (cold phase) La Niña conditions. As one might expect, La Niña hurricane seasons also exhibit significantly more damage along the United States coastal zone than do El Niño hurricane seasons.

| 5 |

Memorable Gulf Hurricanes

Our criteria for choosing which memorable Gulf hurricanes to examine include the life experiences of the coastal inhabitants at each place as well as more objective historical data. Potential factors include storm intensities in terms of the Saffir-Simpson scale (categories one through five); maximum heights of storm surges; loss of life; monetary estimates of property, business, and infrastructure losses; the collective impacts on the socioeconomic functions of communities and cities; and, of course, the personal narratives of the survivors.

The official tropical storm and hurricane data base of the National Hurricane Center shows a total count of 470 tropical storms and hurricanes over the Gulf of Mexico for the 155 years from 1851 through 2005, about three every season on average. Of these, 248, or a little more than half, attained hurricane status over the Gulf, 84 became major hurricanes, and 11 reached the dreaded category five status. This data set is not homogeneous through time, however, and it seems certain that in the nineteenth century some tropical storms and hurricanes have not been identified and that the intensity estimates are questionable. For example, no category five storms have been identified over the entire Atlantic-Caribbean-Gulf waters until 1928, and the first category five storm over the Gulf not until 1935, raising questions now about evaluations of the intensities of historical storms. An alternative view might suggest that this increasing pattern of events represents climate change and global warming. While this could be the case, we feel that changes in methods of hurricane observation over time is the more likely explanation. These methods have changed from land and ship observations in the nineteenth and early twentieth centuries to use of reconnaissance aircraft in the early 1940s. Then, in the late 1960s, satellites, which can observe nearly the entire Atlantic basin in frequent intervals, came into use (note fig. 8.1). Analysis of the data sets suggests that some weaker tropical storms over the Gulf in the nineteenth century were not identified. Of the 50 storms over the Gulf between 1851 and 1875, only 15, or 30 percent, were assigned tropical storm rather than hurricane status. By contrast, of the 69 storms between 1976 and 2000, 34, or almost half, were counted as tropical storms rather than hurricanes.

Major hurricanes over the Gulf (Saffir-Simpson categories three through five) usually have the greatest environmental and socioeconomic impacts along the coasts and tend to be the most memorable ones. There are always exceptions, however, and Tropical Storms Claudette (1979) in the vicinity of Houston, Allison (1989) over southeastern Texas and much of Louisiana, and another Allison (2001) over the same region are very memorable because of record rainfalls and regional flooding. One final example is category one Hurricane Agnes, which came on shore near Apalachicola in 1972, eventually generating catastrophic flooding far to the north in New York, Pennsylvania, Delaware, and Maryland as a tropical storm.

Table 5.1 is a chronological listing of the 84 hurricanes that achieved major hurricane status during their tracks over the Gulf of Mexico from 1851 through 2007, 157 seasons. That is about one every other year on average. The all-inclusive dates are from the NHC data set and span from the storm's initial identification as a tropical depression by the NHC until the system was no longer identified thus. The next column identifies the seasonal storm number as designated by the NHC for 1851–1949, and beginning with 1950 the name given to each storm; some of the same names reappear in different seasons as the storm names are recycled every six years. The next column shows the maximum Saffir-Simpson category that was estimated for the storm track anywhere over the open Atlantic, Caribbean, or Gulf waters. The final column shows our estimates of storm categories at the time of strike, when the centers were no more than 150 miles from the coast and tropical storm or hurricane conditions were usually being experienced at the coastline. There is no obvious long-term trend of major hurricane events over the Gulf from 1851 to 2007. Instead there are short-term clusters of major hurricanes, with four in 1909, two in 2004, and five in 2005, followed by none in 2006 and 2007. Multi-season clusters include 11 major hurricanes in the nine years between 1909 and 1917, 12 in the eleven years between 1960 and 1970, and 11 during the eleven years between 1995 and 2005. At the opposite end of the frequency count, there were no major hurricanes identified during the twelve seasons between 1861 and 1872.

Each of the 84 major hurricanes over the Gulf made landfall along the

Gulf Coast but often weakened near the coastline. Table 5.1 shows that 21 of the 84 major hurricanes over the Gulf made landfall with categories of less than three—twelve at category two, eight at category one, and one, Isidore in 2002, as a tropical storm near Boothville, Louisiana.

Table 5.1. Major Hurricanes over the Gulf of Mexico, 1851—2007

	NHC Storm No./ Storm Name	Max. Saff-Simp. Cat.	Max. Cat. over Gulf	Cat. at Landfall	Loc. at Landfall
1851 Aug. 16–27	4	3	3	3	Panama City Beach, FL
1852 Aug. 19–30	1	3	3	3	Dauphin Is., AL
1855 Sep. 15–17	5	3	3	3	Boothville, LA; Gulfport, MS
1856 Aug. 9–12	1	4	4	4	Isles Dernieres (Morgan City), LA
Aug. 25–Sep. 3	5	3	3	2	Panama City Beach, FL
1860 Aug. 8–16	1	3	3	3	Boothville, LA
1873 Sep. 26–Oct. 10	5	3	3	2	Sanibel Is., FL
1875 Sep. 8–18	3	3	3	3	Port Aransas, TX
1877 Sep. 21–Oct. 5	4	3	3	2	Panama City Beach, FL
1879 Aug. 29–Sep. 2	4	3	3	3	Morgan City, LA
1880 Aug. 4–14	2	4	4	4	South Padre Is., TX
1882 Sep. 2–13	2	3	3	3	Pensacola Beach, FL
1886 Aug. 12–21	5	4	4	4	Indianola, TX
Oct. 8–13	10	3	3	3	Pecan Island, LA
1893 Sep. 27–Oct. 5	10	4	4	4	Grand Isle, LA
1894 Oct. 1–12	5	3	3	3	Panama City Beach, FL
1896 Sep 22–30	4	3	3	3	Cedar Key, FL
1900 Aug. 27–Sep 15	1	4	4	4	Galveston, TX
1906 Oct. 8–23	8	3	3	3	Key West, FL
1909 July 13–22	4	3	3	3	Freeport, TX
Aug. 20–28	6	3	3	3	Lago Morales, Tamaulipas
Sep. 12–22	8	3	3	3	Grand Isle, LA
Oct. 6–13	10	3	3	3	Key West, FL
1910 Oct. 9–23	5	4	4	3	Key West, FL
1915 Aug. 5–23	2	4	4	3	Galveston, TX
Sep. 22–Oct. 1	6	4	4	4	Grand Isle, LA
1916 June 29–July 10	2	3	3	3	Pascagoula, MS
Aug. 12–19	4	3	3	3	Yucatan; South Padre Is., TX

	NHC Storm No./ Storm Name	Max. Saff-Simp. Cat.	Max. Cat. over Gulf	Cat. at Landfall	Loc. at Landfall
1917 Sep. 21–29	3	3	3	3	Pensacola Beach, FL
1919 Sep. 2–15	2	4	4	4	Key West, FL; Port Aransas, TX
1921 Oct. 20–30	6	4	4	3	St. Petersburg, FL
1924 Oct. 14–23	7	3	3	1	Marco Island, FL
1926 Sep. 11–22	6	4	4	3	Pensacola Beach, FL
1932 Aug. 12–18	2	4	4	4	Freeport, TX
1933 Aug. 28–Sep. 5	11	3	3	3	South Padre Island, TX
1935 Aug. 29–Sep. 10	2	5	5	5	Marathon, FL
1942 Aug. 21–31	2	3	3	3	Port O'Connor, TX
1944 Oct. 12–23	11	3	3	3	Key West, FL
1945 June 20–July 1	1	3	3	1	Cedar Key, FL
Aug. 24–29	5	4	4	2	Port O'Connor, TX
1946 Oct. 5–14	5	4	4	1	Sarasota, FL
1948 Sep. 18–25	7	3	3	3	Marco Island, FL
Oct. 3–16	8	4	3	2	Key West, FL
1949 Sep. 27–Oct. 6	10	4	4	4	Freeport, TX
1950 Sep. 1–9	Easy	3	3	3	St. Petersburg, Cedar Key, FL
1951 Aug. 12–23	Charlie	4	4	3	Tampico, Tamaulipas
1953 Sep. 23–28	Florence	3	3	1	Panama City Beach, FL
1955 Sep. 10–20	Hilda	3	3	2	Tampico, Tamaulipas
1957 June 25–29	Audrey	4	4	4	Cameron, LA
1960 Aug. 29–Sep. 14	Donna	5	4	4	Marco Island, FL
Sep. 14–17	Ethel	5	5	1	Boothville, LA
1961 Sep. 3–16	Carla	5	5	4	Yucatan; Port O'Connor, TX
1964 Sep. 28–Oct. 5	Hilda	4	4	3	Morgan City, LA
Oct. 8–17	Isbell	3	3	2	Key West, FL
1965 Aug. 27–Sep. 13	Betsy	4	4	3	Grand Isle, LA
1966 June 4–14	Alma	3	3	2	Apalachicola, FL
Sep. 21–Oct. 11	Inez	4	4	3	Yucatan; Tampico, Tamaulipas
1967 Sep. 5–22	Beulah	5	5	3	South Padre Is., TX
1969 Aug. 14–22	Camille	5	5	5	Boothville, LA
				5	Bay St. Louis, MS
1970 July 31–Aug. 5	Celia	3	3	3	Port O'Connor, TX

(continued)

Table 5.1. (continued)

	NHC Storm No./ Storm Name	Max. Saff-Simp. Cat.	Max. Cat. over Gulf	Cat. at Landfall	Loc. at Landfall
Sep. 8–13	Ella	3	3	3	Lago Morales, Tamaulipas
1974 Aug. 29–Sep. 10	Carmen	4	4	3	Morgan City, LA
1975 Aug. 24–Sep. 1	Caroline	3	3	3	Lago Morales, Tamaulipas
Sep. 13–24	Eloise	3	3	3	Panama City Beach, FL
1977 Aug. 29–Sep. 3	Anita	5	5	4	Lago Morales, Tamaulipas
1979 Aug. 29–Sep. 15	Frederic	4	4	3	Dauphin Island, AL
1980 July 31–Aug. 11	Allen	5	5	3	Yucatan; South Padre Island, TX
1983 Aug. 15–21	Alicia	3	3	3	Galveston, TX
1985 Aug. 28–Sep. 4	Elena	3	3	3	Biloxi, MS
Nov. 15–23	Kate	3	3	2	Panama City Beach, FL
1988 Sep. 8–20	Gilbert	5	4	3	Progreso, Yucatan
				4	Lago Morales, Tamaulipas
1992 Aug. 16–28	Andrew	5	4	4	Sanibel Is., FL
				3	Morgan City, LA
1995 Sep. 27–Oct. 6	Opal	4	4	3	Pensacola Beach, FL
1999 Aug. 18–25	Bret	4	4	3	South Padre Island, TX
2002 Sep. 14–27	Isidore	3	3	3	Progreso, Yucatan
Sep. 21–Oct. 4	Lili	4	4	2	Morgan City, LA
2004 Aug. 9–15	Charlie	4	4	4	Sanibel Is., FL
Sep. 2–24	Ivan	5	5	3	Gulf Shores, AL
2005 July 4–18	Dennis	4	4	3	Navarre Beach, FL
July 11–21	Emily	5	3	3	Boca Madre, Tamaulipas
Aug. 23–31	Katrina	5	5	3	Boothville, LA; Bay St. Louis, MS
Sep. 18–26	Rita	5	5	3	Cameron, LA
Oct. 15–26	Wilma	5	5	3	Marco Island, FL

The 63 major hurricane strikes along the coasts of the Gulf from 1851 to 2007 are listed in table 5.2, arranged around the shores of the Gulf from Key West to the Yucatan. The Florida coastline east and southeast of Panama City for almost 300 miles to the vicinity of Sanibel Island is sheltered from direct strikes by major hurricanes from the east and southeast because of the configuration of the coastline and the breadth of the

Table 5.2. Major Hurricanes Strikes around the Gulf of Mexico, 1851–2007

	Year	No./Name	Max. Cat. in Gulf	Max. Cat. at Time of Strike (NHC/RAM)
Key West, FL	1906	8	3	3
	1909	10	3	3
	1910	5	4	3
	1919	2	4	4
	1944	11	3	3
	1948	8	3	3
Marathon, FL	1935	2	5	5
Marco Is., FL	1948	7	3	3
	1960	Donna	4	4
	2005	Wilma	5	3
Sanibel Is., FL	1992	Andrew	4	4
	2004	Charlie	4	4
St. Petersburg, FL	1921	6	4	3
	1950	Easy	3	3
Cedar Key, FL	1896	4	3	3
	1950	Easy	3	3
Panama City B., FL	1851	4	3	3
	1894	5	3	3
	1975	Eloise	3	3
Navarre Beach, FL	2005	Dennis	4	3
Pensacola B., FL	1882	2	3	3
	1917	3	3	3
	1926	6	4	3
	1995	Opal	4	3
Gulf Shores, AL	2004	Ivan	5	3
Dauphin Island, AL	1852	1	3	3
	1979	Frederic	4	3
Pascagoula, MS	1916	1	3	3
Biloxi, MS	1985	Elena	3	3
Gulfport, MS	1855	5	3	3
Bay St. Louis, MS	1969	Camille	5	5
	2005	Katrina	5	3
Boothville, LA	1855	5	3	3
	1860	1	3	3
	1969	Camille	5	5
	2005	Katrina	5	3

(continued)

Table 5.2. (*continued*)

	Year	No./Name	Max. Cat. in Gulf	Max. Cat. at Time of Strike (NHC/RAM)
Grand Isle, LA	1893	10	4	4
	1909	8	3	3
	1915	6	4	4
	1965	Betsy	4	3
Morgan City, LA	1856	1	4	4
	1879	4	3	3
	1964	Hilda	4	3
	1974	Carmen	4	3
	1992	Andrew	4	3
Pecan Is., LA	1886	10	3	3
Cameron, LA	1957	Audrey	4	4
	2005	Rita	5	3
Galveston, TX	1900	1	4	4
	1915	2	4	4
	1983	Alicia	3	3
Freeport, TX	1909	4	3	3
	1949	10	4	4
	1932	2	4	4
Port O'Connor, TX	1942	2	3	3
	1961	Carla	5	4
	1970	Celia	3	3
Indianola, TX	1886	5	4	4
Port Aransas, TX	1875	3	3	3
	1919	2	4	3
South Padre Is., TX	1880	2	4	4
	1916	4	3	3
	1933	11	3	3
	1967	Beulah	5	3
	1980	Allen	5	3
	1999	Bret	4	3
Boca Madre, Tam.	2005	Emily	3	3
Lago Morales, Tam.	1909	6	3	3
	1970	Ella	3	3
	1975	Caroline	3	3
	1977	Anita	5	4
	1988	Gilbert	4	4

	Year	No./Name	Max. Cat. in Gulf	Max. Cat. at Time of Strike (NHC/RAM)
Tampico, Tam.	1951	Charlie	4	3
	1966	Inez	4	3
Progreso, Yucatan	1966	Inez	4	4
	1988	Gilbert	4	3
	2002	Isidore	3	3
Rio Lagartos, Yucatan	1966	Inez	4	3
	1980	Allen	5	4
	1988	Gilbert	4	3
	2002	Isidore	3	3
Isla Holbox,				
Quintana Roo, Mex.	1916	4	3	3
	1961	Carla	5	3
	1966	Inez	4	3
	1980	Allen	5	5
	1988	Gilbert	4	3
	2002	Isidore	3	3

northern and central Florida Peninsula. For this extended coastline during this period of record, there have been only three major hurricane strikes, all three coming on shore at Cedar Key. The southwestern coast of Florida from Key West northward to Sanibel Island has experienced more than its fair share, with three major hurricane strikes at both Key West and Marco Island.

The northern Gulf Coast, extending about 400 miles eastward from Freeport, Texas, to Panama City Beach, Florida, has been hit often with a dismal history of devastating major hurricane strikes. Of the 63 major hurricane strikes around the shores of the Gulf from 1851 to 2007, 36—more than half of all the strikes—have assaulted the coastline from East Texas to the Florida Panhandle. The most frequent strike events have occurred from Morgan City, Louisiana, eastward about 100 miles in a straight line to Gulfport, Mississippi, where 15 major hurricane strikes between 1855 and 2007, one every ten years on average, have resulted in catastrophic socioeconomic losses and environmental damage. There have been five strikes at Morgan City, four at nearby Grand Isle, and four at Pensacola Beach in the Florida Panhandle.

For the entire 156 years there have been no major hurricane strikes at landfall along the Gulf coasts of Mexico south of Tampico with the exception of the Yucatan Peninsula, and, indeed, the first strike from Tampico northward to the border of the United States after 1850 did not occur until 1909. Table 5.2 does show, however, that there have been eight major hurricane strikes along the nearly 300-mile coast of the Mexican state of Tamaulipas between 1909 and 2005, a coast that has always been sparsely populated with the exception of the city of Tampico on the southern border with the state of Veracruz. Immediately to the north of the Mexican border, at South Padre Island, there have also been six additional major hurricane strikes between 1880 and 1999, making a grand total of 14 strikes from Tampico northward for about 350 miles, including South Padre Island. This extended coast is especially vulnerable to major hurricanes sweeping westward over the open Gulf, with minimal opportunities for weakening as the storms approach the coast.

We turn now to reviews of the tracks and impacts of some of these most memorable major hurricane events around the shores of the Gulf of Mexico.

Cheniere Caminada Hurricane, October 1893

This unexpected tragic hurricane attained tropical storm status over the northwest Caribbean Sea on September 27, 1893 (see fig. 5.1, the underlying data for all storm track maps are from the NHC). The next day, intensification brought the cyclone to category two status before landfall near Cancun on the Caribbean coast of the Yucatan. After emerging from the Yucatan, the storm intensified rapidly to category four over the northern Gulf before making landfall on October 2 at the fishing village of Cheniere Caminada, just west of Grand Isle, Louisiana. After landfall, the storm maintained hurricane category one status during a northeasterly track across coastal Mississippi to southwestern Alabama.

Because of the limited meteorological observation and communication technologies of the time, the hurricane, with its sudden sixteen-foot storm surge, surprised and overwhelmed the 1,500 inhabitants of Cheniere Caminada. About 800 of the villagers, more than half of the population, drowned, and the total loss of life in southeastern Louisiana, coastal Mississippi, and Alabama was about 2,000 people. One survivor on a makeshift raft was found eight days later 100 miles to the southeast, near the South Pass of the Mississippi River. Fig. 5.2 shows the only home

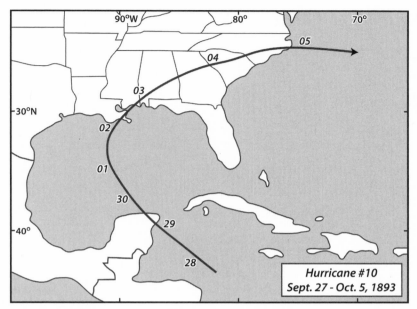

Fig. 5.1. Track of Cheniere Caminada Hurricane of 1893.

Fig. 5.2. The only home to survive the Cheniere Caminada Hurricane in coastal Louisiana in 1893.

in the village that survived, though it is severely damaged. Severe coastal damage and drownings also occurred in coastal Mississippi and as far east as Pensacola, Florida. The orange, rice, and sugarcane crops were severely damaged or destroyed. Damage has been estimated at $100 million in 2007 dollars, and would be much higher because of coastal developments along the central Gulf Coast if the storm took place today.

The Galveston Hurricane, September 1900

This hurricane is discussed in chapter 1.

Southeast Louisiana Hurricane, September 1915

This hurricane developed over the eastern Caribbean Sea southeast of Puerto Rico on September 22, 1915 (see fig. 5.3). The storm then passed over the Yucatan Channel as a category three hurricane on September 27 and strengthened to category four only 170 miles southeast of New Orleans on September 28 before rapidly weakening to a category three as it

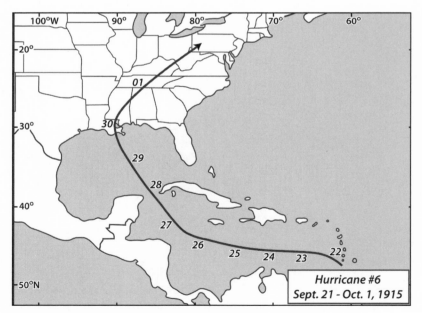

Fig. 5.3. Track of the Southeastern Louisiana Hurricane of 1915.

Fig. 5.4. The Poland Station Streetcar Barn in New Orleans following the 1915 Hurricane.

made landfall near Grand Isle, Louisiana. Nevertheless, Plaquemines and St. Bernard Parishes were struck by storm surges up to thirteen feet, and more than one hundred people drowned in Plaquemines Parish alone and twenty-one in New Orleans. New Orleans experienced severe wind damage, including the loss of more than half of the church steeples in the city, the complete collapse of the First Presbyterian Church, and destruction of the New Orleans streetcar barn (see fig. 5.4). The hurricane also destroyed the remnants of a Filipino village built on stilts in the marsh of St. Bernard Parish on the shores of Lake Borgne. The village was initially settled by Filipino sailors who fled Spanish galleons in the late eighteenth century.

From 1901, the highest official Weather Bureau/National Weather Service sustained-wind record for New Orleans of 86 miles per hour was recorded during this storm—a New Orleans record that stood until Katrina in 2005. It should be noted that during Katrina the official wind instrumentation failed, but an unofficial top wind speed of 95 miles per hour at Michoud in eastern New Orleans has been accepted by most professional meteorologists.

The Great Labor Day Hurricane, Florida Keys, September 1935

This small but intense hurricane, which struck a narrow band of islands of the Florida Keys on September 2, 1935, was the first category five hurricane landfall on record in the United States (see fig. 5.5). To this day, it is the strongest storm ever to make landfall in the United States, and the third most intense Atlantic-Caribbean-Gulf hurricane known, after Wilma in 2005 and Gilbert in 1988. The storm was first identified with tropical storm status on Thursday, August 29, east of the Bahamas and about 800 miles east of Miami. By Saturday the storm had drifted westward about 300 miles but still remained classified by the Weather Bureau as a tropical storm rather than a hurricane. On Sunday, September 1, however, the storm began to intensify rapidly, with category one status by 7 a.m., category two by 7 p.m., category three by 1 a.m. on Labor Day morning, category four by 7 a.m., and category five by 7 p.m., just before landfall at Matecumbe and Long Keys, very close to the village of Islamorada, about thirty miles southwest of Key Largo and about 80 miles northeast of Key West. The hurricane then curved northward over the Gulf and made a second landfall at Cedar Key, Florida, as a category two hurricane on Wednesday, September 4.

The post-1929 economic depression continued to weigh heavily on the country in 1935, and about 680 World War I veterans were encamped at Islamorada working on construction of an overseas highway across the Upper Keys for the Federal Emergency Relief Administration (FERA). About 300 of these veterans had gone on Labor Day leave to a baseball

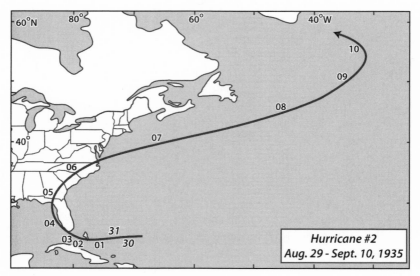

Fig. 5.5. Track of the Labor Day Hurricane of 1935.

Fig. 5.6. This 11-car train was washed off its tracks by the Labor Day Hurricane of 1935. Photo taken on September 5, 1935.

game in Miami, with the remaining 380 veterans at their low-lying camps in the projected path of the hurricane on Labor Day afternoon. The only escape link between the Keys was the single-track Florida Overseas Railroad, a unit of the Florida East Coast Railroad Company. A special rescue train consisting of a locomotive and eleven cars left Miami just before 5 p.m., finally arriving at Islamorada about 8:20 p.m., minutes before a massive storm surge wave swept over the Key, ripping all of the cars except for the locomotive off the track and demolishing nearly every structure in the village.

About 440 people drowned—280 veterans and 160 local inhabitants. The storm surge was estimated at eighteen to twenty feet, and railroad tracks thirty feet above the sea were washed off the railroad viaducts. An 11-car train was sent down to the Keys to rescue World War I veterans housed there in a rehabilitation camp. Although the train reached the site, the storm swept it off the tracks (fig. 5.6). The hurricane was very small, and the swath of total destruction along the Keys was only about twenty miles. Nearly thirty-five miles of railroad track, the only link between Key West and the mainland, was destroyed, never to be rebuilt. Ernest Hemingway visited the desolation of the Keys two days after the storm and, on September 17, published in the *New Masses* magazine, "Who Murdered the Veterans? A First-Hand Report on the Florida Hurricane." Subsequent investigations of the questions of responsibility eventually absolved federal, local, and railroad officials from the perspective of forecast technologies of the time, with the hurricane track much farther north than predicted and with no expectation of the rapid record-breaking intensification.

As mentioned above, the hurricane curved northward over the Gulf after landfall on the Florida Keys and struck Cedar Key as a category two storm, with five additional deaths there from winds and flooding. The railroad was never rebuilt, but many of the railroad bridges and trestles that survived the hurricane were recycled for the nearly 100-mile Overseas Highway from Key Largo to Key West that was opened to vehicle traffic in 1938.

Hurricane Audrey, June 1957

On June 25, 1957, the residents of low-lying Cameron Parish, adjacent to the Gulf in southwestern Louisiana, were probably not too alarmed when the Weather Bureau reported the initial development of Audrey as a tropical storm over the Bay of Campeche about 500 miles south of the town of Cameron, the parish seat of government (see fig. 5.7). After all, severe hurricanes almost never developed so early in the season (June), and furthermore, there had only been three minor hurricane strikes with little damage at Cameron during the previous fifty-six seasons of the twentieth century. Later the same day, nevertheless, Audrey evolved into a category two hurricane, moving northward toward the Cameron Parish coast.

During the following day, June 26, Audrey continued on its direct

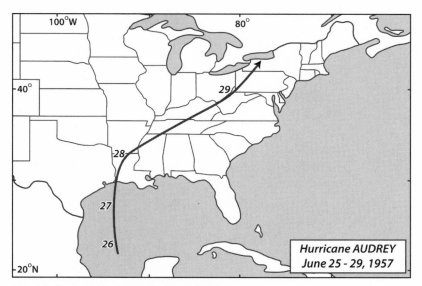

Fig. 5.7. Track of Hurricane Audrey in 1957.

path toward the parish. With its intensity wavering between a minimum hurricane, category one, and the category two of the previous day, many residents elected to remain in their homes overnight, since Audrey was predicted to make landfall in Cameron Parish as a minimal hurricane during the afternoon of June 27. The decision to stay was undoubtedly based on local experiences over many decades with tropical storms and weak hurricanes. Overnight, however, Audrey not only increased its forward motion toward the coast but also intensified unexpectedly.

Early in the morning, residents awoke to a full-blown category four storm with winds in excess of 140 miles per hour, and a twelve-foot storm surge that swept as much as 25 miles inland across coastal marshes, with escape to distant higher ground impossible. About 525 people drowned, more than one-third under nine years old, and countless others were never accounted for. Along the immediate coast from Cameron east to Grand Cheniere, 60 to 80 percent of the homes and buildings were heavily damaged or destroyed, and the hurricane's estimated $1 billion loss in 2007 dollars makes Audrey the most destructive hurricane of its time since the 1938 hurricane in New England. After landfall, Audrey continued on toward the northeast and became a very destructive extratropical storm across the lower Mississippi and Ohio valleys.

Hurricane Betsy, September 1965

"Billion Dollar Betsy" was the most costly hurricane in the history of the United States at its time, with the total cost exceeding $1 billion for the first time—in 2007 dollars, that would be $6.5 billion. Betsy was first identified as a hurricane northeast of Puerto Rico on August 30, 1965. Its track was toward the northwest, and by September 4 category four Betsy was 400 miles east of Daytona Beach, threatening the coasts of the Carolinas (see fig. 5.8). However, Betsy turned around and executed a great loop to the south, making landfall as a category three hurricane at Key Largo in southern Florida on September 8. Betsy then emerged over the eastern Gulf, intensifying briefly to category five only 140 miles southeast of New Orleans before making a second landfall as a category three storm at Grand Isle, Louisiana, a barrier island fifty miles south of New Orleans. Betsy destroyed almost all of the residential and commercial structures and the facilities for offshore recreational fishing in Grand Isle.

The strongest sustained wind recorded in New Orleans was only 69 miles per hour, not quite hurricane force, but our hurricane strike

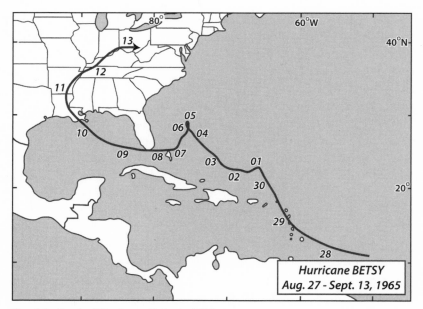

Fig. 5.8. Track of Hurricane Betsy in 1965.

Fig. 5.9. Flooding in New Orleans resulting from Hurricane Betsy in 1965.

model indicates a category one strike at New Orleans. At the city the storm surge was about ten feet, and many of the protective levees were breached, flooding much of the city, especially the Lower Ninth Ward and Chalmette, for ten days (fig. 5.9). Up to 250,000 people had to be evacuated, with about 150,000 housing units flooded. Seventy-five to eighty residents drowned. Hurricane Betsy also destroyed Manila Village, another Filipino settlement in Barataria Bay, established in the late nineteenth century for catching and drying shrimp.

President Lyndon B. Johnson flew to New Orleans the following day and quickly produced a federal outline for aid and rebuilding infrastructure. As a result of Betsy, the Corps of Engineers was authorized to rebuild the levees to sixteen feet—high enough, it was believed, to protect the city from flooding associated with another category three hurricane.

Hurricane Beulah, September 1967

After several days as a tropical depression over the central Atlantic, Beulah was upgraded to a tropical storm in the vicinity of St. Lucia over the extreme eastern Caribbean Sea on September 7, 1967 (see fig. 5.10). Two days later, Beulah exploded to a category four major hurricane about 125 miles south of Puerto Rico. Beulah continued on a westward track across the northern Caribbean but was downgraded to tropical storm status on September 12 after interacting with the mountainous terrain of Hispaniola. Beulah regained strength over the western Caribbean and came on shore as a category two storm on September 17 near Cozumel on the Caribbean coast of the Yucatan.

After emerging over the Bay of Campeche the next day, Beulah exploded to a very large category five hurricane with its center only about 100 miles southeast of South Padre Island. Beulah made landfall September 20 as a category four storm near the mouth of the Rio Grande and close to the southern tip of South Padre Island.

Very effective evacuation orders extended 150 miles northward from

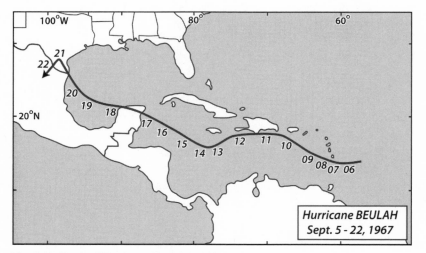

Fig. 5.10. Track of Hurricane Beulah in 1967.

South Padre Island through Port Aransas to Rockport, and there were only fifteen deaths associated with Beulah in Texas. An eighteen-foot storm surge caused thirty-one temporary cuts through Padre Island, and the popular beach resort at South Padre Island was heavily damaged. Over inland Texas, Beulah generated 115 mostly small tornadoes and extremely heavy rains, up to twenty-seven inches at Beeville, with record-breaking floods along the Nueces River. There were heavy losses across the irrigated Lower Rio Grande agricultural region, especially to the ruby-red grapefruit crop. Total storm losses in Texas are estimated at $8 billion, adjusted to 2007.

Hurricane Camille, August 1969

Because of its record high winds and storm surge, Camille became the benchmark hurricane strike along the northern Gulf Coast until Hurricane Katrina in 2005. Camille was first identified as a tropical storm over the western Caribbean near Grand Cayman on August 14, 1969 (see fig. 5.11). One day later, Camille struck the western tip of Cuba as a category three hurricane, moving into the Gulf slightly diminished in intensity to a category two status. On August 17, just south of the lower delta of the Mississippi River, the storm's intensity peaked at category five with maximum winds estimated between 190 and 205 miles per hour. Camille made landfall a few hours later near Venice, Louisiana—the strongest landfall in recorded United States and possibly worldwide, history, a record apparently still standing in 2007. Camille crossed the narrow strips of natural levee land on both sides of the Mississippi River in Plaquemines Parish and struck Bay St. Louis on the Mississippi Gulf Coast, still a category five hurricane a few hours later. The inhabitants of metropolitan New Orleans narrowly escaped with minimal property losses on the much weaker left side of the hurricane circulation.

Camille totally destroyed lower Plaquemines Parish below Buras, normally the domicile of several thousand people. There was almost no loss of life because of ample warnings of the hurricane threat and strictly enforced evacuation orders. Several Weather Bureau personnel barely survived in their elevated facilities for launching radiosondes to obtain upper-air weather data in Boothville. The storm surge covered exit doors, and they had to escape through a ceiling hatch. Along the Mississippi Coast, almost all buildings from the Louisiana border to Biloxi and up to

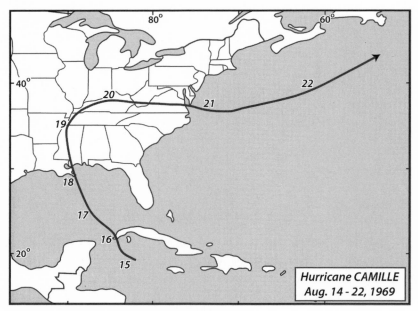

Fig. 5.11. Track of Hurricane Camille in 1969.

three miles inland were destroyed, much of it by the record-high storm surge of 22.6 feet at Bay St. Louis. Much has been written about a "hurricane party" on the third floor of the Richelieu Manor Apartments adjacent to the beach in Pass Christian; there were only two survivors of twenty-three who elected to stay behind at the beachfront complex instead of evacuating inland (fig. 5.12). In total, more than 140 people drowned along the Gulf Coast during Camille, almost all in Mississippi. Most of the survivors along the immediate coast had to endure several days without communication with the outside world or fresh food and water, as the access roads were blocked by thousands of fallen pine trees and the multiple bridges were washed out.

After landfall on the Gulf Coast, Camille continued north to northern Mississippi for 150 miles as a hurricane. The remnants of the storm then moved northeastward, bringing unexpected torrential rains of twelve to twenty inches along the mountains of the James River basin in West Virginia and Virginia. The resultant mountainside mudflows and flash floods caught inhabitants of the valley bottoms by complete surprise. More than 150 people drowned—a larger total than along the Gulf

Fig. 5.12. The Richelieu Manor Apartments in Pass Christian, Mississippi, before and after the 23-foot storm surge associated with Hurricane Camille in 1969.

Coast, where there had been ample warnings. Total losses for Camille amounted to about $9 billion (adjusted to 2007 values), exceeding even the losses associated with Hurricane Betsy in 1965 and thus making Camille the most costly hurricane, in southeastern Louisiana and the Mississippi Gulf Coast up to that time.

Hurricane Anita, August and September 1977

Hurricane Anita was classified as a hurricane for only three days, but Anita reached category five status before coming on shore with winds estimated at 175 miles per hour. Anita initially hit the very sparsely settled coast of the Mexican state of Tamaulipas, in the vicinity of Soto la Marina, about 145 miles south of Brownsville, Texas, and eighty miles north of Tampico (see fig. 5.13). Anita was first designated as a tropical depression on August 29, about 200 miles southeast of New Orleans. Less than twenty-four hours later, Anita was upgraded to tropical storm status and then to category one hurricane status. By noon on September 1, Anita had intensified to category three status, and then to category five status just six hours later, coming on shore in Tamaulipas as a category five on September 2. The next day, the much weakened storm circulation reached the eastern Pacific as a tropical depression.

While Anita was over the western Gulf, upper-air steering currents were initially weak, and Anita was forecasted to make landfall near the Louisiana-Texas border. The approximately 20,000 residents of Cameron Parish, Louisiana, remembering the fate of many who did not evacuate ahead of Hurricane Audrey in 1957, left the vulnerable low-lying coastal areas. The next day Anita threatened the southern Texas coast between

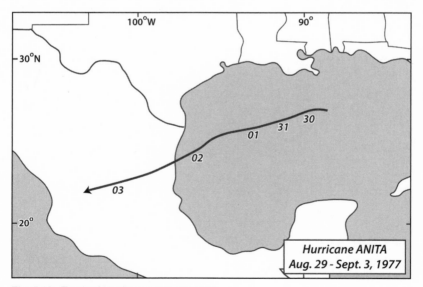

Fig. 5.13. Track of Hurricane Anita in 1977.

Corpus Christi and Brownsville, but a developing high pressure ridge to the northwest forced Anita further south to the upper Mexican coast instead.

Socioeconomic impacts of this category five strike were minimal because few residents in Tamaulipas were in harm's way. The death toll was officially only ten, but 25,000 people were estimated homeless because of heavy rains, flooding, and mudslides. Eighteen inches of rain were recorded at an observation site near Anita's landfall. As Anita gathered strength over the central Gulf, a storm surge of two feet was reported at Grand Isle, Louisiana, three inches of rain were recorded at nearby Galliano, and Texas highway 87 along the beaches of High Island between Port Arthur and Galveston was closed because of migration of beach sand associated with overwashes.

Hurricane Allen, August 1980

Allen was a Cape Verde storm that developed to hurricane status in the mid-Atlantic, becoming a major hurricane on August 3 (see fig. 5.14). The following day Allen achieved category four status just south of St. Lucia, and then attained category five status on August 5 over the eastern Caribbean, about 200 miles south of Puerto Rico. Allen threaded its way across the Caribbean, alternating between categories four and five before brushing by the northern coast of the Yucatan Peninsula; the eye never touched

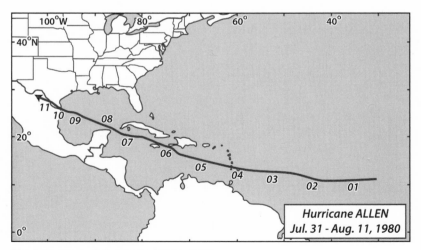

Fig. 5.14. Track of Hurricane Allen in 1980.

land during this long journey across the Caribbean. On August 9, Allen again intensified to category five only 125 miles southeast of South Padre Island. Luckily, Allen weakened rapidly as it approached the coast, making landfall in the vicinity of the southern tip of South Padre Island as a category three hurricane. Allen logged more time as a category five storm than any other Atlantic hurricane and equaled Hurricane Camille's record maximum wind speed of 190 miles per hour.

Damages associated with Allen amounted to only $2.7 billion in 2007 dollars. There were 274 deaths overall, including 24 in the United States. Seventeen of those deaths occurred in Louisiana when a helicopter crashed during evacuation of an offshore oil platform.

Hurricane Gilbert, September 1988

Hurricane Gilbert was the most intense hurricane over the entire Atlantic-Caribbean-Gulf until Hurricane Wilma in 2005. Gilbert began as a cluster of thunderstorms on September 3 off the west coast of Africa, making it a Cape Verde Hurricane. The disturbed weather pattern held together as it moved westward across the tropical Atlantic. Five days later, on September 8, it was designated by the Hurricane Center as Tropical Depression 12, nearly 400 miles east of Barbados. On September 9, the circulation achieved tropical storm status in the vicinity of Martinique and was named Gilbert (see fig. 5.15).

On September 11, on a very consistent track toward the northwest, Gilbert became a hurricane and intensified rapidly to major hurricane status, category three, south of Puerto Rico and the Dominican Republic. On September 12, Gilbert raked the southern coast of Jamaica with category three winds and heavy flooding rains. Rains were recorded up to twenty-seven inches, resulting in forty-five deaths and much property and infrastructure damage, estimated at $4 billion in 1988 dollars ($6.9 billion in 2007 dollars). The following day, Gilbert swept past Grand Cayman Island, again as a category three hurricane, but without inflicting major damage. Later the same day, south of the western tip of Cuba, Gilbert exploded to category five status, with a central atmospheric pressure of 888 millibars—at that time the lowest pressure ever measured in the Western Hemisphere—and top winds estimated at 185 miles per hour. Gilbert maintained category five status until it crashed ashore at Cozumel on the Caribbean coast of the Yucatan Peninsula on September 14.

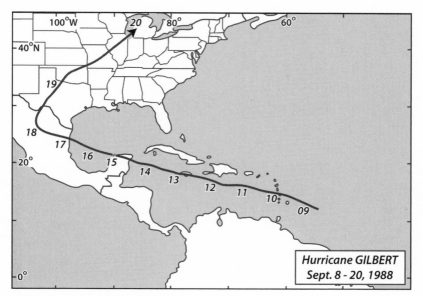

Fig. 5.15. Track of Hurricane Gilbert in 1988.

Destruction to this international resort island was extremely severe, but no lives were lost because of early predictions and warnings of the potential danger of a category five strike on the island. Nevertheless, on the Yucatan Peninsula 60,000 homes were destroyed and 83 vessels sunk in adjacent waters.

On its track of about 200 miles over the Yucatan, Gilbert weakened to category two status. Over the very warm waters of the Gulf, Gilbert regained major hurricane status (category four) before coming on shore on September 16 at a small fishing village, La Pesca, in the Mexican state of Tamaulipas. This location was about 150 miles south of Brownsville, Texas, very close to where Hurricane Anita came on shore as a category five storm eleven years earlier in 1977. The coastal areas there were sparsely settled, resulting in minimal loss of life and property damage. However, the rapidly weakening circulation of Gilbert, combining with a ten-thousand-foot mountain barrier, was still capable of unleashing record rainfalls and devastating floods around Monterrey, where more than one hundred people died when five buses transporting evacuees were overturned by raging floodwaters on September 17. There were 225 hurricane-related deaths in Mexico. The circulation remnants of Gilbert eventually turned northward as a tropical depression across western Texas on September 18 with significant rainfalls over the steppe landscapes accompanied by twenty-nine small tornadoes.

Altogether, during its six-day track with hurricane status over the Caribbean and the Gulf, Gilbert was responsible for 341 lost lives and about $8.7 billion in property losses (in 2007 dollars), much of this in Jamaica and Cozumel.

Two Tropical Storms Allison, June 1989 and June 2001

These two tropical storms, twelve years apart, will remain memorable to many of the residents of southeastern Texas and adjacent parts of Louisiana for enormous rainfall totals and consequent massive urban flooding in Houston and Baton Rouge. This was the result of slow-moving tropical circulations that barely met tropical storm status. Allison in 1989 produced almost thirty inches of rain in five days at Winnfield, Louisiana; eleven drownings in Texas, Louisiana, and Mississippi; and losses of almost $1 billion in 2007 dollars. Tropical Storm Allison in 2001 made landfall very close to the 1989 Allison storm, produced nearly forty inches of rainfall locally near downtown Houston, was responsible for forty-one drownings, of which twenty-three were in Texas, and resulted in about $6 billion in economic losses.

On June 24, 1989, an evolving tropical disturbance was identified by the National Hurricane Center over the southwestern Gulf about 75 miles southeast of Corpus Christi, Texas, and about 175 miles southwest of Galveston (see fig. 5.16a). The disturbance was the result of the merging of a weak tropical wave moving westward over the Gulf with the remnants of Pacific Hurricane Cosme, which had crossed central Mexico and emerged out over the Gulf. The following day, the disturbance intensified to tropical storm status but was still positioned about 50 miles southeast of Corpus Christi. Allison came on shore just west of Freeport on June 26 as a weak tropical storm circulation, but with unusually heavy rainfalls. Soon after landfall, Allison was downgraded to a tropical disturbance, drifting slowly over Houston on June 27. On June 28, the storm moved ever so slowly northeastward to the vicinity of Lufkin, Texas, and then stalled and eventually turned back toward the southwest. On June 30, the poorly defined center was again over Houston before moving out once again toward the northeast on July 1 and eventually all the way to the East

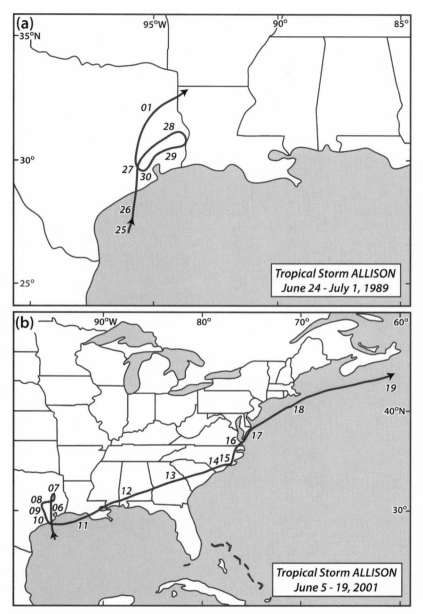

Fig. 5.16. Tracks of Tropical Storms Allison in a) 1989 and b) 2001.

Coast, where more heavy rainfalls generated flooding and transportation disruptions between New York and Washington.

Twelve years later, on June 4, 2001, a tropical disturbance was positioned over the northwestern Gulf relatively close to the position of Tropical Storm Allison in 1989 (see fig. 5.16b). One day later, June 5, the storm system intensified to tropical storm status, another Tropical Storm Allison. The tropical storm, about 150 miles south of Freeport, made landfall later the same day at Freeport with maximum winds of 50 miles per hour. The storm was poorly predicted, and locals across the Houston metropolitan area were mostly caught off guard by the heavy rains. Allison drifted slowly over Houston as a tropical depression, and by June 8 had reached the vicinity of Lufkin. Very much like the Allison of 1989, the 2001 storm drifted back near Freeport on June 10, emerging over the Gulf for a second round of tropical storm status. Allison then drifted eastward along the northern Gulf Coast, with a second landfall near Morgan City, Louisiana. Allison moved slowly toward the northeast again as a tropical depression over Mississippi, Alabama, Georgia, South Carolina, and North Carolina, emerging over the Atlantic near Norfolk, Virginia, reaching tropical storm status for a third time off Atlantic City.

Three interrelated rounds of flooding rainfalls characterized the 2001 storm during its extended assault on Houston. The first, during the afternoon and early evening of Tuesday, June 5, during the landfall near Freeport, produced eight to twelve inches of rain over southern and eastern sections of Houston and Harris County. The second wave, during the morning of Thursday, June 7, with five to twelve inches of rain across mostly southern areas of Houston and Harris County, fell while Allison was meandering slowly toward the northeast across the metropolitan region (see fig. 7.5).

The final onslaught, during the late afternoon and evening of Friday, June 8, and overnight into Saturday morning, was associated with the return of Allison as a tropical depression southwestward from the vicinity of Lufkin back offshore at Freeport. These flooding rains were the most devastating of the three events, with four inches of rain in an hour and a maximum total of twenty-six inches in one rain gauge in the northeastern section of Houston. Nearly all of the freeways and major roads in and around downtown Houston were flooded and impassible overnight and on Saturday. The most destructive damage, however, was to the various downtown medical center complexes, collectively among the largest in the world, with flooded basements and first floors, power outages, and thousands of patients in need of evacuation by small boats and helicopters. Combined losses of the medical centers came to about $2 billion ($2.3 billion in 2007 dollars). Many downtown office towers, as well as the symphony, opera, and theater buildings, were also damaged by floodwaters, with total storm losses in the Houston metropolitan region estimated at about $7 billion (in 2007 dollars). These are amazing losses for such a weak tropical meteorological system—hardly identifiable on the daily weather maps for North America.

Hurricane Andrew, August 1992

Hurricane Andrew will never be forgotten for its terrible assault on the southeastern Florida coast, especially between Miami and Homestead, but this hurricane is also especially memorable for its subsequent Gulf of Mexico strikes on the west coast of Florida and then, several days later, on southern Louisiana. Andrew was first identified as a tropical storm over the Atlantic midway between Africa and the Caribbean on August 17, 1992 (see fig. 5.17). It was not until August 22 that the National Hurricane Center upgraded Andrew to a category one hurricane about 650

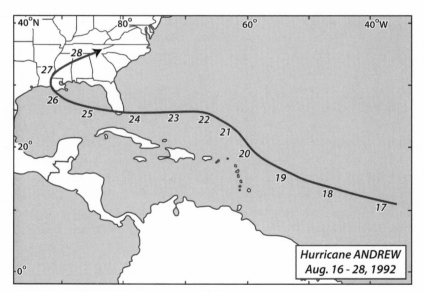

Fig. 5.17. Track of Hurricane Andrew in 1992.

Fig. 5.18. Tornado damage in LaPlace, Louisiana, associated with Hurricane Andrew in 1992.

miles east of Nassau in the Bahamas. Little more than twenty-four hours later, however, Andrew had rapidly intensified to a category five hurricane with a trajectory directly toward southern Florida and less than twenty-four hours until landfall.

Andrew came on shore near Homestead during early-morning darkness on August 24 as a category five hurricane with winds of 170 miles per hour over a narrow, fifteen-mile-wide totally devastated band extending inland from the coast to the Everglades. The National Hurricane Center in Coral Gables was significantly damaged but remained operational during the storm. About 80,000 dwellings were severely damaged or destroyed, and in south Florida alone, Andrew became the most costly storm to date in American history, about $39 billion in 2007 dollars, with about $1.5 billion of the total occurring later in Louisiana. Most of the destruction in Florida was due to wind rather than storm surge, and the resultant loss of life was limited to forty-three people there. In 2004 National Hurricane scientists reclassified Hurricane Andrew as a category five storm at landfall near Homestead, making Andrew only the third category five hurricane to make landfall in the United States, along with the Florida Keys storm in 1935 and Camille in 1969.

Andrew crossed the Everglades in less than four hours, emerging over the Gulf a few miles south of Marco Island as a category four hurricane. Andrew then curved slowly toward the northwest and then to the north, maintaining category four status until just before landfall on August 26 about 20 miles west of Morgan City, Louisiana, a region of coastal marshes with few inhabitants. Andrew then moved rather slowly toward the northeast, passing 20 miles west of Baton Rouge as a tropical storm with a peak wind gust at the airport of 69 miles per hour. Nevertheless, storm winds over Baton Rouge brought down thousands of large oak and pecan trees as well as power lines. Bob Muller remembers being without electrical power for eight sweltering days. There were only four deaths in Louisiana, two of them from a tornado in LaPlace, near New Orleans (fig. 5.18). It was estimated that about 1.25 million people evacuated their homes in Louisiana, especially around New Orleans, another quarter of a million left southeastern Texas, and about one million people had earlier vacated southern Florida.

The 2004 and 2005 Seasons

The dreadful memories of these consecutive seasons along the coasts of the Gulf of Mexico will remain for years. In 2004 there were two major hurricane strikes: the first, Hurricane Charley, at Charlotte Harbor, just to the north of Sanibel Island in southwestern Florida; and the second, Hurricane Ivan, in the western Florida Panhandle and Alabama (see fig. 5.19). There were also four tropical storm strikes along the west coast of the Florida Peninsula; two of these four were dissipating stages of major hurricanes that made initial landfalls along the east coast of the peninsula.

Hurricane Charley was first classified as a tropical storm over the eastern Caribbean Sea on August 10. One day later it was reclassified as a minimal hurricane moving northwestward south of Jamaica, and the following day Charley barely missed Grand Cayman as a category two. On August 13 Charley made landfall in southwestern Cuba as a category three storm, passing just west of Havana on the same day with category two status. Passing over the southeastern Gulf quickly, Charley intensified rapidly to make landfall at Captiva Island near Port Charlotte, less than twenty miles northwest of Fort Myers as a category four storm and less than twenty-four hours after Tropical Storm Bonnie made landfall in the Florida Panhandle. Charley was the strongest landfall in the United States since Hurricane Andrew just south of Miami in 1992. Charley

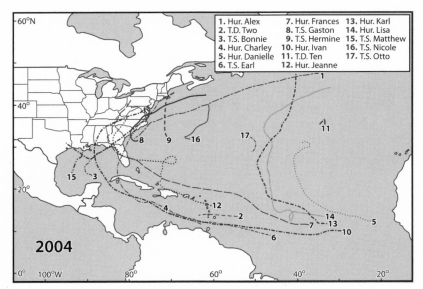

1. Hur. Alex	**7.** Hur. Frances	**13.** Hur. Karl
2. T.D. Two	**8.** T.S. Gaston	**14.** Hur. Lisa
3. T.S. Bonnie	**9.** T.S. Hermine	**15.** T.S. Matthew
4. Hur. Charley	**10.** Hur. Ivan	**16.** T.S. Nicole
5. Hur. Danielle	**11.** T.D. Ten	**17.** T.S. Otto
6. T.S. Earl	**12.** Hur. Jeanne	

Fig. 5.19. Tropical storm and hurricane tracks for the 2004 season.

continued to move rapidly toward the northeast, bringing category two wind conditions to Orlando and a category one strike at Daytona Beach on the east coast. Hurricane Charley is given credit for two additional "brush-by" category one strikes in South Carolina and North Carolina.

The forecasting of the landfall and intensity of Charley as it approached the southwest Florida coast was especially difficult because of the orientation of the coast relative to the storm track. Charley was approaching the coast from the south-southwest, and the coast orientation is essentially north-south. A very small change in direction could result in a landfall many miles away. Indeed, five hours prior to landfall, Charley was predicted to strike the St. Petersburg and Tampa areas as a category two hurricane, but after a slight shift in track two hours before landfall, the predicted strike was moved to Port Charlotte, about sixty miles to the south, and the intensity upgraded to category four! After massive evacuations in the St. Petersburg-Tampa metropolitan region, not much of significance happened there, but many of the residents around Charlotte Harbor had not evacuated and were shocked by the catastrophic and deadly category four strike of Charley. Altogether, about two million residents were urged to evacuate, but it is estimated that only half did so.

Because of Charley's small size, the surge near landfall at Port Charlotte was only about six feet, with a maximum surge of more than ten feet to the south near Naples. Port Charlotte and Punta Gorda, both near landfall, suffered severe wind damage and were leveled. Inland toward Orlando there was a ten-mile-wide swath of severe destruction along Charley's track toward the East Coast. Agricultural losses in the citrus groves amounted to more than $200 million. Total losses for the United States were estimated at $15 billion, making Charley the fourth most costly hurricane to date. Nevertheless, only ten people lost their lives because of the hurricane.

Hurricane Ivan is an example of a classic Cape Verde Hurricane that originated in disturbed tropical weather off the west coast of Africa. Ivan was first identified as a tropical storm on September 3 and as a minimal hurricane more than one thousand miles east of the Windward Islands on September 5. Ivan intensified to category three, major hurricane status, the same day, at only ten degrees of north latitude; it is extremely rare for a hurricane to develop and intensify at such a low latitude.

On September 7 Ivan crossed over Grenada, leaving the island suffering from major damage. Two days later Ivan further intensified to category five status just north of Aruba, again setting a new record for a category five hurricane so far south over Atlantic or Caribbean waters. Ivan oscillated back and forth between category four and five two more times as it stormed along very close to Jamaica on September 10 and again close to Grand Cayman on September 11. Ivan reached its maximum intensity with winds of 170 miles per hour in the vicinity of Grand Cayman. After crossing the Yucatan Channel and moving over the Gulf, Ivan maintained category four status until landfall on September 16 at Orange Beach, Alabama, just west of Pensacola, Florida, as a category three hurricane.

As Ivan approached the northern Gulf Coast, the track was so uncertain that evacuations were urged or mandatory for the Florida Keys and from the central Florida Panhandle to southeastern Louisiana. About half of the population of the New Orleans metropolitan region left in a poorly organized evacuation, with many motorists taking twelve hours or more to reach Baton Rouge, a drive that normally requires little more than an hour. Nevertheless, the failed appearance of major hurricane Ivan at New Orleans undoubtedly led to a spate of professional and media analyses pronouncing New Orleans as the greatest natural disaster waiting to happen in the United States. Ivan's storm surge was extremely destructive east of the landfall all the way to Panama City Beach, Florida, with nearly total waterfront destruction at Pensacola and Perdido Key. In the United States there were eighty-four deaths associated with the hurricane and about $13 billion in losses. These figures were in addition to the prior widespread destruction and loss of life on Grenada, Jamaica, and the Cayman Islands.

After landfall near the Alabama-Florida border, Ivan weakened rapidly to tropical depression status over northern Alabama. The tropical depression circulation continued on in a great clockwise loop that crossed over Maryland and Delaware, then southward over the western Atlantic, then westbound across southern Florida, and finally out over the Gulf again. Ivan then briefly regained tropical storm status and came on shore again along the northern Gulf Coast, this second time at the Louisiana-Texas border west of Cameron, Louisiana, before finally dissipating over southeastern Texas. Ivan set records for being a tropical depression, tropical storm, or hurricane for twenty-two days from September 2 to September 24.

The 2005 season was even more deadly. There were three major hur-

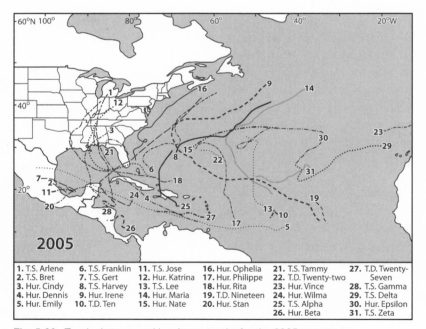

1. T.S. Arlene	6. T.S. Franklin	11. T.S. Jose	16. Hur. Ophelia	21. T.S. Tammy	27. T.D. Twenty-
2. T.S. Bret	7. T.S. Gert	12. Hur. Katrina	17. Hur. Philippe	22. T.D. Twenty-two	Seven
3. Hur. Cindy	8. T.S. Harvey	13. T.S. Lee	18. Hur. Rita	23. Hur. Vince	28. T.S. Gamma
4. Hur. Dennis	9. Hur. Irene	14. Hur. Maria	19. T.D. Nineteen	24. Hur. Wilma	29. T.S. Delta
5. Hur. Emily	10. T.D. Ten	15. Hur. Nate	20. Hur. Stan	25. T.S. Alpha	30. Hur. Epsilon
				26. Hur. Beta	31. T.S. Zeta

Fig. 5.20. Tropical storm and hurricane tracks for the 2005 season.

ricane strikes along the northern Gulf Coast: Hurricane Dennis, a follow-up from the previous season in the western Florida Panhandle; Hurricane Katrina in Louisiana and Mississippi; and Hurricane Rita in western Louisiana and eastern Texas (see fig. 5.20). Hurricane Emily struck the northern Mexican coast south of Brownsville, Texas. In addition, there were two other hurricanes—Cindy, in Louisiana; and Stan, over the Bay of Campeche and the adjacent Mexican coast—and four tropical storms. The season began early with Arlene, which became a tropical storm strike on June 8 in the vicinity of Pensacola Beach, Florida. The tracks of the first five storms of the season were partially or entirely over the Gulf, and indeed, eight of the first eleven storms were over the Gulf at some time.

Hurricane Cindy attained tropical storm status on July 5 over the central Gulf and made landfall as a minimal hurricane the following day at Grand Isle, Louisiana. After crossing the lower delta of the Mississippi River, Cindy made a second landfall as a tropical storm in the vicinity of Biloxi, Mississippi. This minimal storm generated eight inches of rain at Galliano in southern Louisiana and thirty-three small tornadoes over the southeastern states. Altogether, damage estimates amounted to $320 million. This storm is interesting in that its original maximum intensity noted by the NHC was tropical storm strength, but it was upgraded to a minimal hurricane several weeks after the fact. This upgrade had serious insurance claim implications.

Hurricane Dennis was the first of five major hurricanes to make landfall around the Gulf during the 2005 season. Dennis reached category four status south of Cuba over the western Caribbean before crossing western Cuba. In Cuba, Dennis was credited with destroying 15,000 homes and damaging another 120,000 homes, resulting in more than $2 billion in losses.

Departing Cuba, Dennis was downgraded to category one but regenerated to category four in less than twenty-four hours over the very warm loop current in the Gulf. Dennis weakened a bit as the storm approached the Florida Panhandle but nevertheless was credited with a category three strike at Navarre Beach, Florida, on July 10, about thirty miles east of the landfall of Hurricane Ivan one year earlier in 2004. Dennis was a much smaller storm than Ivan, but the storm surge was destructive to residential and commercial developments along the barrier islands of the Panhandle, resulting in damage estimates of about $2.2 billion.

Hurricane Emily originated as a tropical wave off the African coast on

July 6 and was declared a hurricane on July 14 in the vicinity of Grenada. It became the earliest category five storm on record for the entire Atlantic Basin while over the Caribbean south of Hispaniola on July 16. Emily continued on a west-northwest track south of Jamaica and made landfall as a category four hurricane at Cozumel, where there was heavy damage to the tourist industry infrastructure but very limited loss of life because of effective evacuation procedures. Emily continued over the Gulf to a second landfall as a category three hurricane at La Pesca in the Mexican state of Tamaulipas, a sparsely populated coast. There were only six deaths, and damage estimates were about $400 million in Mexico, most of it in the vicinity of Cancun.

Hurricane Katrina, another category five hurricane, battered southeastern Louisiana and the Mississippi Gulf Coast on August 29. See chapter 2.

Hurricane Rita, yet another category five hurricane, developed initially at tropical storm status to the west of the Turks and Caicos Islands on September 18. Rita moved westward and attained hurricane status over the Florida Straits on September 20. Then, within a twenty-four-hour span, it exploded to a category five while tracking over the very warm loop current in the Gulf, with maximum sustained winds estimated at 180 miles per hour. Rita turned northward toward the central Gulf Coast, with landfall first thought to be close to New Orleans again and then later the vicinity of Galveston and Houston. Ultimately, Rita made landfall in between as a category three hurricane just east of the Texas-Louisiana border near Johnson Bayou. In Louisiana the coastal fishing and oil-service town of Cameron was completely destroyed, and, in Texas, the city of Port Arthur was heavily damaged.

Along the way, Rita caused the most massive evacuations for a single storm in American history, in total estimated at about three million people. More than 300,000 people evacuated southern Florida, with mandatory evacuations of all of the Florida Keys from Key West to Key Largo. Evacuations were ordered for all communities along the Louisiana coast including New Orleans, where heavily damaged levees from Hurricane Katrina were breeched again and some areas of the city flooded one more time. Several days in advance of landfall, with an evacuation plan in place, residents of Galveston Island were ordered to leave. One day later, the mayor of Houston urged all residents there to evacuate, resulting in massive gridlock on the highways to the west and north. The "drive" to Dallas was reported to take between twenty-four and thirty-six hours, and to Austin, twelve to eighteen hours, all in sweltering midday temperatures approaching one hundred degrees Fahrenheit!

Nevertheless, there were only seven deaths attributed directly to the storm, though there were a considerable number of additional deaths related one way or another to Rita. The most widely publicized were the deaths of twenty-three evacuees in a bus accident southeast of Dallas. Total damages were estimated at $10 billion. Along the five-hundred-mile stretch of Gulf Coast from near Galveston in the west to Panama City Beach, Florida, in the east, there remained hardly an undamaged structure because of the category three strikes from east to west by Dennis at Navarre Beach, Florida (July 2005), Ivan at Orange Beach, Alabama (September 2004), Katrina at Buras, Louisiana (August 2005) and Bay St. Louis, Mississippi (August 2005), and Rita at Johnson Bayou, Louisiana (September 2005).

Hurricane Wilma, the final category five hurricane of 2005, was identified officially as a tropical storm over the western Caribbean Sea on October 17. Drifting slowly to the west, Wilma developed into a record-breaking category five hurricane southwest of Jamaica on October 19. Two days later, on October 21, Wilma struck the Caribbean coast of the Yucatan, including the tourist resorts of Cancun and Cozumel on October 22. There was severe damage to resort establishments especially at Cancun, where storm waves reached the third floors of beach hotels. An incredible storm rainfall of sixty-four inches was recorded at Isla Mujeres, a small Caribbean island less than ten miles east of Cancun. Damage estimates in Mexico amounted to about $7.5 billion.

Wilma then turned abruptly to the northeast over the Gulf toward Florida, coming on shore as a category three hurricane in the vicinity of Sanibel Island, causing extensive damage from Fort Myers south to Marco Island. Wilma moved rapidly across the Florida Peninsula and was still capable of significant destruction on the east coast, especially around Fort Lauderdale. Damage to the power grid was so extensive that some customers remained without electricity for two to three weeks. Two storm surges swept across the lower Florida Keys, the first from the south as Wilma approached Florida from the west, and the second from the north as Wilma swept across the southern Florida mainland. There were five deaths in Florida attributed directly to the storm and about $21 billion in damages.

In chapter 6, we look at the history of hurricane strikes from 1901 to 2007 at ten coastal towns and cities representative of hurricane climates around the Gulf to provide some perspective of potential hurricane hazards along the different coasts.

| 6 |

HURRICANE HISTORIES

In this chapter, we investigate the geography and history of hurricane strikes at ten cities around the Gulf of Mexico, counterclockwise from Key West to Progreso, Mexico, on the Yucatan Peninsula (see fig. 6.1). The record begins with 1901 except for New Orleans, where we go back to 1851. The selection of locations is designed to illustrate the geographical pattern of coastal reaches, some with much greater, others much lesser, threats of hurricane strikes, and the time span reveals clusters of years with more frequent strikes as contrasted to runs of years with little activity at each place.

We begin with the very active history on the small island of Key West,

Fig. 6.1. Locations of the cities around the Gulf Basin examined in this chapter.

the southernmost city of the forty-eight contiguous states, with the Atlantic waters of the Straits of Florida to the south and east and the waters of the Gulf of Mexico to the north and west. Moving northward along the eastern shore of the Gulf, we come to St. Petersburg. Here the strike record clearly reflects the mitigating effects of the 150-mile-wide Florida Peninsula on the hurricanes approaching and crossing Florida from southeast to northwest, as well as the vulnerability to hurricanes traveling from south to north over the eastern Gulf, with almost no opportunities for weakening due to the configuration of the coastline.

Our third stop is along the very northeastern corner of the Gulf at the small city of Apalachicola, where the coast is much more open to hurricanes sweeping northward over the eastern Gulf, but where more land to the east and north apparently help reduce the chances for major hurricane strikes. There have been no major hurricane strikes at Apalachicola since 1901, but, we should expect to witness a major strike eventually.

Pensacola Beach is located more or less in the middle section of the northern Gulf Coast. This coast is open to strikes from the southeast, south, and southwest, and the frequency of strikes, and even of major hurricane strikes, is not much lower than at Key West.

New Orleans, the foremost seaport of the entire Gulf until the rise of Houston in the twentieth century, would have a strike record somewhat similar to that of Pensacola Beach were it not for the city's location farther inland from the open coast. New Orleans is sheltered from the onslaught of hurricane strikes by the extended delta of the Mississippi River to the south and southeast, though the coastal marshes and wetlands are currently being lost to the Gulf at alarming rates. The hurricane strike records at Gulfport, Boothville, and Morgan City are similar to the history of strike frequencies at Pensacola Beach, and their hurricane strike records are included in this discussion of New Orleans. Because of the historical significance of New Orleans, we have extended the strike record there back to 1851.

Galveston is positioned at the western end of the northern Gulf Coast, and the frequency and intensity of hurricane strikes there during the 107 years are similar to the patterns at Pensacola Beach and Boothville.

Port Aransas (near Corpus Christi) is positioned along the northwestern coast of the Gulf. The frequency of hurricane strikes drops off there because there are no opportunities for hurricanes to approach from the southwest. Although there have been no major hurricane strikes in the record for Port Aransas, major hurricanes approaching from the east and southeast have occurred along this coast, as is evident from the major hurricane strike record at Port O'Connor and South Padre Island in table 5.2.

The strike record at Tampico is representative of the record for northern Mexico, along the western coasts of the Gulf. This coastal region is vulnerable to hurricanes approaching from the east and southeast, with occasional major strikes due to great hurricanes coming from the Caribbean and crossing the central Gulf.

Southward to Veracruz and beyond around the Bay of Campeche, the southwest coast of the Gulf enjoys a very low frequency of hurricane strikes, even though the bay is thought of as a breeding region for development of tropical storms and hurricanes. At Veracruz there have been only two hurricane strikes beginning with 1901 and no major strikes. Indeed, at one of our additional strike analysis sites, Coatzacoalcos, there have not been any hurricane strikes during this period.

Along the western and northern coasts of the Yucatan Peninsula, the seaport of Progreso serves as our final focus. There the frequency and intensity of hurricane strikes, and major hurricane strikes, increase again. However, the intensity of some strikes is weakened because hurricanes from the Caribbean often first make landfall on the eastern coast of the Yucatan Peninsula before crossing the peninsula to strike Progreso.

Key West, Florida

Key West is at the western end of the Overseas Highway, which extends from the Florida mainland just north of Key Largo southwestward to Key West for 113 miles, mostly over water. The waters to the north and west of the Keys are Florida Bay and the Gulf of Mexico, and to the south and east, the Straits of Florida and the Atlantic Ocean. Key West is the southernmost city of the continental United States. The island is about four miles long and two miles wide, with a maximum elevation of only eighteen feet.

The resident population is about 25,000, and a significant number of tourists throughout the year add to the total. Evacuation before an approaching hurricane is tedious and difficult, especially because most of the Overseas Highway consists of two-lane roads and bridges, and evacuation extends to the resident population and tourists on each of the islands for 113 miles to Key Largo. Given the number of potential evacuees and the destructive winds, surges, and blinding rains well in advance of landfall, the time needed for safe evacuation must be a minimum of twenty-four hours and more than forty-eight hours in unusual situations.

All of the tropical storm and hurricane strikes at Key West are considered Gulf of Mexico storms, even though many arrive at Key West from the east and southeast off of Atlantic waters. All of these storms cross over the Keys and immediately move over Gulf waters; hence, they are Gulf of Mexico hurricanes. Beginning with 1901, there have been sixteen tropical storm and twenty hurricane strikes, seven of which are classified as major hurricane strikes. The tracks of the hurricane strikes are shown in figure 6.2. Three clusters of hurricane strikes testify to the persistence of atmospheric circulation patterns controlling hurricane tracks: 1906–1910, three hurricanes in 48 months; 1947–1950, four hurricanes in 36 months; and 1964–1966, four hurricanes in 24 months. As of this writing, there have been no major hurricane strikes since 1966, and there were no hurricanes for twenty years between 1967 and 1986. In terms of environmental impact, hurricane storm surges tend to be lower along the Keys than along the mainland coasts because surge waters generally pile up against the coasts but sweep by offshore islands.

October 18, 1906: storm 8; category three strike at Key West

First identified as a tropical storm over the southern Caribbean 750 miles southeast of the Grand Cayman on October 8, this storm made its first landfall as a category three hurricane the next day, north of Bluefields, Nicaragua. After a tortuous journey over the jungles and mountains of Central America, the hurricane emerged one week later on October 16 over the northwestern Caribbean, off the Yucatan Peninsula. One day later it was a category three hurricane racing northeastward over the Yucatan Channel, and on October 18 it passed thirty miles southeast of Key West before another landfall in the middle Florida Keys. This hurricane was small in size, however, and Key West sustained only minor damage. Tragically, a houseboat on Long Key that was housing laborers working on the construction of the Overseas Railroad to Key West lost its moorings and broke up at sea, resulting in the loss of about 125 men. The bi-

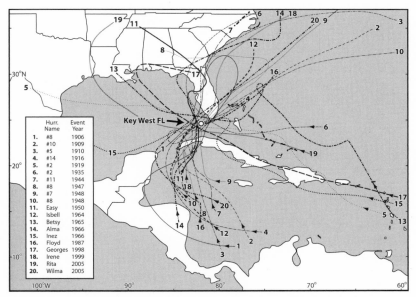

Fig. 6.2. Hurricanes having impact at Key West, Florida, 1901–2007.

zarre track of this hurricane later included an arrival on October 20 as a category one hurricane fifty miles southeast of Charleston, South Carolina, followed by a complete about-face and a return past Key West again as a tropical storm on October 21.

October 11, 1909: storm 10; category three strike at Key West

This storm originated as a tropical storm on October 6 over the southern Caribbean Sea 650 miles south of Grand Cayman. By October 9 it was over the Isle of Youth, Cuba, with a category three intensity. Two days later the hurricane, still a category three, passed just to the southeast of Key West on its way to landfall near Flamingo in Everglades National Park. The storm surge covered much of Key West, with four hundred buildings destroyed, three hundred boats lost, and twelve railroad workers drowned.

October 17, 1910: storm 5; category three strike at Key West

This was the third hurricane strike at Key West in four years. Each of these hurricanes formed initially over the southern or western Caribbean. This one, on October 11 about 550 miles south of Grand Cayman, tracked largely northward to Key West. The hurricane attained category three and four status during a slow, two-day counterclockwise loop over the Gulf west of Cuba from October 14 to 16, followed by a rapid thrust toward Key West. The category four storm passed less than fifty miles to the west of Key West on October 17 before landfall on the southwestern coast of Florida near Fort Myers. Much of Key West was again inundated by the storm surge, but damage was less than during the 1909 hurricane. Probably because of good storm predictions, there was no loss of life at Key West.

November 15, 1916: storm 14; category one strike at Key West

This was another Key West hurricane that formed over the southwestern Caribbean, 750 miles southeast of Grand Cayman on November 11. The minimal hurricane passed over the Yucatan Channel as a category one storm on November 14, with the eye of the hurricane passing very close to Key West on the following day with little damage at Key West.

September 10, 1919: storm 2; category four strike at Key West

This great hurricane over extreme southern Florida was first observed as a tropical storm in the vicinity of Martinique on September 2. By September 8 it had intensified to a major hurricane over the central Bahamas and to category four one day later, only fifty miles southeast of Key West, with the eye passing to the southwest about twenty miles away. The very large hurricane continued westward across the Gulf to make ultimate landfall near Corpus Christi, Texas, where hundreds of people drowned.

There was again major damage and destruction in Key West associated with the storm surge, unusually high waves, and hurricane-force winds for more than twenty-four hours. Surprisingly, only three people lost their lives in Key West. Nevertheless, several large passenger vessels sank during the hurricane in the Florida Straits, with four to five hundred people losing their lives.

September 2, 1935: storm 2; Great Labor Day Hurricane; category two strike at Key West

This deadly hurricane that battered the middle Keys is considered to be the first historic category five hurricane strike in the United States. The hurricane passed about fifty miles northeast of Key West over Florida Bay and the Gulf as a category four storm. A summary of the tragic events is presented in chapter 5.

October 18, 1944: hurricane 11; category three strike at Key West

During this very active season, storm 11 was identified as a hurricane over the western Caribbean three hundred miles south of Grand Cayman on October 13. By October 17 it was a category three just south of the Isle of Youth. The following day the eye passed just west of Havana, where there was a massive storm surge and much coastal destruction and loss of life. Still category three, the hurricane passed about fifty miles west of Key West on its way to landfall near Sarasota, with only minor damage at Key West.

October 11, 1947: storm 8; category one strike at Key West

This tropical storm was first recognized over the western Caribbean three hundred miles south of Grand Cayman on October 9. It moved northward quickly, however, and strengthened to hurricane status on October 11, only fifty miles southwest of Key West. Later the same day the minimal hurricane gave Key West a glancing blow as it passed just twenty-five miles to the west, on its track to landfall near Marco Island.

September 21, 1948: storm 7; category three strike at Key West

On September 18 a tropical storm was identified over the western Caribbean just 175 miles southeast of Grand Cayman. One day later the storm was a category one hurricane 250 miles southeast of the Isle of Youth, and the following day, the same storm was an ugly category three only about twenty-five miles west of Havana. The next day, September 21, the category three hurricane eye passed less than ten miles east of Key West, where a six-foot storm surge flooded many structures. Landfall was near Marco Island.

October 5, 1948: storm 8; category three strike at Key West

Only two weeks later another major hurricane originated over the western Caribbean, followed a track somewhat similar to that of the previous hurricane, and passed again very close to Key West as a category three hurricane. This storm was first identified three hundred miles south of Grand Cayman off the eastern cape of Honduras on October 3, intensifying to a minimum hurricane the next day about 175 miles south of the western tip of Cuba. On October 5, the category four hurricane roared across Havana before passing twenty-five miles east of Key West as a category three storm, with surprisingly less damage to Key West than occurred during the previous hurricane.

September 3, 1950: Hurricane Easy; category one strike at Key West

This fourth hurricane to strike Key West in three years again followed the familiar track from the western Caribbean northward to Key West. Tropical Storm Easy was recognized two hundred miles southwest of Grand Cayman on September 1. The next day it was a hurricane, and one day later Hurricane Easy passed twenty-five miles west of Key West as a minimal hurricane with very minor damage. Nevertheless, Hurricane Easy drifted northward to make two complete loops on September 4 and 5 in the vicinity of Cedar Key, almost completely destroying that small isolated fishing village with two hurricane landfalls within less than thirty-six hours.

October 14, 1964: Hurricane Isbell; category two strike at Key West

Isbell was another Key West hurricane that developed over the western Caribbean Sea. Isbell appeared first as a tropical storm about 275 miles west-northwest of Grand Cayman on October 13, and it intensified rapidly on its track to the north, passing only twenty-five miles to the northwest of Key West as a category three hurricane on October 14. Because of the rapid forward speed of Isbell, there was only minimal damage at Key West and at landfall later in the day at Everglades City.

September 8, 1965: Hurricane Betsy; category one strike at Key West

Hurricane Betsy was a major hurricane strike that initially hit Key Largo and then later Grand Isle, Louisiana, near New Orleans. In between, the eye of category three Betsy passed less than fifty miles to the north of Key West on September 8. Damage at Key West was minimal, except for a number of boats that sank. Hurricane Betsy is discussed in chapter 5 as a memorable hurricane.

June 8, 1966: Hurricane Alma; category two strike at Key West

Alma was another of the hurricanes that developed over the western Caribbean and tracked northward to Florida, in this case 250 miles southwest of Grand Cayman. Alma made landfall near Apalachicola on June 9, and its track history is described in that section. One day earlier, on June 8, the center of Alma passed to the west of Key West as a category three hurricane, but again with minimal damage to the city.

October 4, 1966: Hurricane Inez; category one strike at Key West

Inez was identified as a tropical storm over the central Atlantic on September 24 and developed to a major hurricane over the northeastern Caribbean between Guadeloupe and Puerto Rico on September 28. Inez was extremely damaging and deadly over the Dominican Republic, Haiti, and Cuba, but the track over mountainous sections of these large islands weakened its circulation to tropical storm status by the time it reached the Florida Straits on October 4. The following day Inez, on a track toward the west, passed about twenty miles to the south of Key West as a minimal hurricane. Four times in three hurricane seasons, Key West escaped serious damage and destruction on the fringes of these storms. Inez continued across the Gulf to Mexico, making a major hurricane landfall north of Tampico.

October 12, 1987: Hurricane Floyd; category one strike at Key West

Floyd was first observed as a tropical storm over the western Caribbean, 250 miles south of Grand Cayman on October 10. After tracking to the north, Floyd achieved minimal hurricane status on October 12 only seventy-five miles southwest of Key West. Although Floyd passed over Key West, the minimal hurricane did not cause any significant damage there.

September 25, 1998: Hurricane Georges; category two strike at Key West

On September 20 Hurricane Georges attained category five status about 650 miles east of Antigua, but its track over the mountains of Puerto Rico, Hispaniola, and Cuba disrupted the circulation so much that Georges was only a category one or two hurricane as it crossed the Florida Straits toward Key West. On September 25 Georges passed as close as twenty-five miles to the southwest of Key West as a category two strike there, on the track that led eventually to the Mississippi Gulf Coast. The storm surge flooded much of Key West, and about 1,500 dwellings were damaged significantly.

October 15, 1999: Hurricane Irene; category one strike at Key West

Irene reached tropical storm status over the western Caribbean about 250 miles south of Grand Cayman on October 13. Much like Hurricane Floyd in 1987, Irene achieved minimal hurricane status only seventy miles southwest of Key West on October 15. Irene is remembered in Key West mostly for urban street flooding associated with heavy tropical rains.

September 20, 2005: Hurricane Rita; category two hurricane at Key West

Rita appeared first as a tropical storm on September 18 over the southern Bahamas, passing Key West about forty miles to the south as a category two hurricane, on its way to a devastating onslaught later over southwestern Louisiana and southeastern Texas (see chapter 5). Its storm surge flooded sections of the Overseas Highway east of Key West, but storm damage, fortunately, was light.

October 24, 2005: Hurricane Wilma; category two strike at Key West

On October 17 Wilma developed as a tropical storm over the western Caribbean about two hundred miles southeast of Grand Cayman. Wilma intensified as it drifted slowly toward the west and southwest, and two days later reached category five status about two hundred miles southwest of Grand Cayman. Wilma made its initial landfall at Cozumel on the Yucatan Peninsula on October 22 as a category three hurricane, before turning toward the northeast and South Florida. On October 24 Wilma passed sixty-five miles to the northwest of Key West as a category three storm and made landfall the same day at Cape Romano, only ten miles south of Marco Island.

The mandatory evacuation ordered for Key West was said to be ignored by about 80 percent of the residents; they had to cope with an eight-foot storm surge that flooded about 60 percent of the houses in Key West. Wilma is also discussed as a memorable hurricane in chapter 5.

St. Petersburg, Florida

St. Petersburg is located on the southern tip of a peninsula, with the Gulf of Mexico to the west and large Tampa Bay to the east. It is often called the Sunshine City in recognition of only the handful of days a year with heavy cloud cover, persistent rains, or lack of sunshine. The Tampa Metropolitan Region, including St. Petersburg, has about 2.7 million inhabitants. It is especially populated to the south, where there are well-developed barrier islands, the most famous being Longboat Key, with beautiful subtropical landscaping that appears to have escaped the ravages of tropical storms and hurricanes.

At Key West, we counted twenty hurricane strikes, including seven major strikes, during the 107 years between 1901 and 2007. Moving north along the coast toward St. Petersburg, we count sixteen hurricane strikes, seven major, at Marco Island, and twelve hurricane strikes, three major, at Sanibel Island farther to the north. At St. Petersburg there have been only four hurricane strikes, two of which were classified as major. Figure 6.3 shows the tracks of these four hurricanes.

Clearly, the geography of hurricane strikes along the Gulf Coast shows an especially steep decline from very high frequencies at southwestern Florida coasts to very infrequent strikes along the middle coast.

October 25, 1921: storm 6; category three strike at St. Petersburg

Tropical storm 6 was identified on October 20 over the Caribbean Sea 550 miles east-southeast of Grand Cayman. Three days later, the storm was over the Yucatan Channel, 550 miles south-southwest of St. Peters-

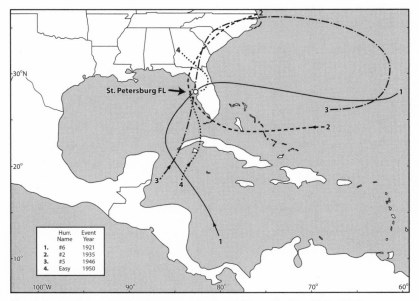

Fig. 6.3. Hurricanes having impact at St. Petersburg, Florida, 1901–2007.

burg, now a major hurricane. It made landfall near St. Petersburg on October 25, still as a major hurricane.

The ten- to twelve-foot storm surge devastated the Tampa commercial waterfront and grand homes near the water's edge. In Ybor City, a Latin neighborhood of Tampa, at least five hundred homes were wrecked. The waterfront at St. Petersburg was also heavily damaged, with a greater proportion of the damage in the region attributed to water rather than wind. The hurricane came at the beginnings of the Florida land boom on the peninsula's west coast. At the time, there were nationwide reports of the hurricane and its damage, which spawned concerns by businesses and industry that this would limit further development. Needless to say, these concerns helped to hasten an extremely rapid cleanup of the region.

September 4, 1935: storm 2; category one strike at St. Petersburg

The Great Labor Day Hurricane was the first category five landfall in the United States ever recorded. Its category five strike in the Florida Keys is discussed in chapter 5 as a memorable hurricane. After crossing the Florida mainland, the hurricane represented a category one strike at St. Petersburg. Damages in the Tampa-St. Petersburg area were minor, but the hurricane arrived the same day as a scheduled local election in Tampa, when it was hoped that the electorate would turn out to overthrow a particularly corrupt city administration.

October 8, 1946: storm 5; category one strike at St. Petersburg

After being observed as a tropical storm over the western Caribbean off Belize on October 5, this system became a hurricane the following day over the Yucatan Channel. It intensified to a category four hurricane only 175 miles south-southwest of St. Petersburg. Despite the potential threat, this hurricane weakened dramatically as it approached landfall near Bradenton, and losses were minor there and at St. Petersburg. This hurricane represented a category one strike at St. Petersburg on October 8.

September 5, 1950: Hurricane Easy; category three strike at St. Petersburg

This storm was first identified as a tropical storm on September 1, about two hundred miles west of Grand Cayman. Easy was upgraded to a hurricane on the following day when it was located just south of Isle of Youth, Cuba. On September 3, Easy exploded to major hurricane status only seventy-five miles southwest of St. Petersburg. Easy was a rare category three strike at St. Petersburg, but the hurricane did not make official landfall until two days later in the vicinity of Cedar Key, about one hundred miles farther north.

Easy's storm surge of about six feet did some damage to shoreline facilities in the Tampa-St. Petersburg area, including the loss of about forty beachfront homes from Sarasota north to Clearwater. Easy moved toward the north very slowly, coming to a stall just off the coast near Cedar Key before coming ashore just south of Cedar Key. The hurricane then executed a tight counterclockwise loop, moving offshore and then returning again for a second landfall about thirty miles south of Cedar Key. The results of the double strike at Cedar Key were more than eighteen hours of continuous hurricane-force winds, record rainfalls of up to thirty inches, and the near-complete destruction of all the buildings in this small commercial and resort fishing village. Incidentally, our hurricane strike record for Cedar Key shows just three hurricane strikes since 1900, with only Easy in 1950 as a major hurricane strike.

Apalachicola, Florida

The small city of Apalachicola is located at the mouth of the Chattahoochee-Apalachicola river system that drains most of western Georgia, including the Atlanta metro area and some of eastern Alabama. Apalachicola's regional position is on the shore of the northeastern corner of the Gulf of Mexico, a coastal section that is developed much less for tourism than for beaches far to the south, closer to Clearwater and St. Petersburg, and beaches to the west, beginning with Panama City Beach. Apalachicola is sheltered from the direct onslaughts of tropical storms and hurricanes that approach from the east and southeast because of the breadth of the Florida Peninsula. Because of this configuration of land and water, there have been even fewer hurricane strikes east and south of Apalachicola, all the way to just north of St. Petersburg. During the 107 years between 1901 and 2007, there have been ten hurricane strikes but no major strikes at Apalachicola. The tracks are shown in figure 6.4.

September 13, 1903: storm 3; category one strike at Apalachicola

This third storm of the season was first identified as a tropical storm on September 9 over the southeastern Bahama Islands. The storm made its first landfall as a category one hurricane at Miami two days later. After

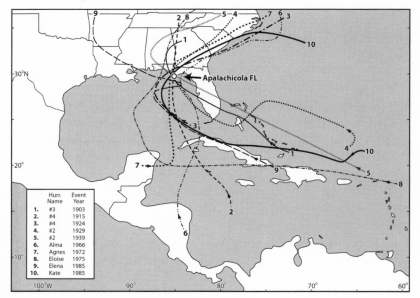

Fig. 6.4. Hurricanes having impact at Apalachicola, Florida, 1901–2007.

exiting the Florida Peninsula near Sarasota, the storm had a second land-fall, as a minimal hurricane, on September 13, at Cape San Blas, about twenty-five miles west-southwest of Apalachicola. The storm surge was reported to be as high as ten feet at Apalachicola, with regional losses to the agricultural, forestry, and fishing industries.

September 4, 1915: storm 4; category one strike at Apalachicola

Twelve years later, Apalachicola was struck by another minimal hurricane. This storm, however, originated over the western Caribbean south of Jamaica on August 31, and after crossing the western tip of Cuba, moved northward as a category one hurricane, coming on shore again on September 4 in the vicinity of Cape San Blas. The sponge fleet was cited for significant losses along with expected damages to waterfront property.

September 15, 1924: storm 4; category one strike at Apalachicola

This minimal hurricane evolved over the eastern Gulf of Mexico as a tropical storm about one hundred miles southwest of Key West on September 12. The storm was categorized as a hurricane on the following day. The track led directly to landfall as a category one hurricane near

Port St. Joe, just a few miles to the west of Cape San Blas. Storm 4 was the third of these minimal hurricanes to make landfall in roughly the same coastal area in twenty-one years. Only minor damage was reported at Apalachicola.

September 30, 1929: storm 2; category one strike at Apalachicola

This erratic hurricane was first identified as a tropical storm over the open Atlantic Ocean about four hundred miles north of Puerto Rico on September 21. By September 26 the storm had intensified to category four and was positioned only two hundred miles east-southeast of Miami. Weakening to category two, the storm crossed Key Largo two days later and moved over the eastern Gulf. On September 30, the storm was classified as category one, only forty miles south of Apalachicola. Landfall was again near Cape San Blas with minor damage around Apalachicola.

August 13, 1939: storm 2; category one strike at Apalachicola

The track of this hurricane was somewhat similar to that of the 1929 hurricane. On August 7 a tropical storm was detected just north of the Dominican Republic. The storm made its initial landfall as a category one hurricane about twenty miles north of Palm Beach on August 11. The following day the now-weakened tropical storm emerged over the Gulf, northwest of Cedar Key, making its second landfall one day later as a category one hurricane at Apalachicola. Again, damage was minimal and limited mostly to waterfront properties.

June 9, 1966: Hurricane Alma; category one strike at Apalachicola

Alma became a tropical storm and category one hurricane on June 6 over the western Caribbean Sea between Honduras and Cuba. Alma moved more or less straight north and, after crossing Cuba, attained major hurricane status briefly about fifty miles southwest of Key West on June 8. The weakening hurricane then remained offshore but close to the west coast of Florida, with a five-foot storm surge. Storm winds were responsible for moderate coastal damage all the way from Key West to Apalachicola, especially in the St. Petersburg-Tampa area and at Cedar Key, where the bridge to the mainland washed out. Alma made landfall as a minimal hurricane at St. George Island twenty miles east-southeast of Apalachicola on June 9. Because Apalachicola was to the left of the storm track, there was little damage in the town.

June 19, 1972: Hurricane Agnes; category one strike at Apalachicola

The track of Hurricane Agnes from the western tip of Cuba northward to the Florida Panhandle was similar to Alma's, except farther to the west. Agnes originated as a tropical storm on June 16 over the Yucatan Channel. Two days later Agnes became a minimal hurricane over the Gulf, northwest of Cuba and about 450 miles south of Apalachicola. The next day Agnes reached landfall at Cape San Blas, twenty-five miles southwest of Apalachicola, weakening and close to the threshold between tropical storm and category one hurricane. Over the Gulf Agnes was a very large hurricane, and the five- to eight-foot storm surges on the east side of its track again resulted in significant coastal damage and destruction from Key West to Panama City and caused several deadly tornadoes over southern Florida. The surge was especially destructive from Apalachicola eastward to around the shores of Apalachee Bay, where many waterfront buildings were heavily damaged or washed away.

After landfall Agnes continued northeastward as a tropical storm to release record rainfalls and floods in much of the Middle Atlantic States of Virginia, Maryland, Delaware, New Jersey, Pennsylvania, and New York, becoming for its time one of the most costly storms in terms of property and infrastructure. Even though wind damage is often minor, tropical storms occasionally generate enormous rainfalls and flooding, much like the two tropical storms Allison in Texas and Louisiana discussed in chapter five.

September 23, 1975: Hurricane Eloise; category one strike at Apalachicola

Hurricane Eloise was a category one strike at Apalachicola and Pensacola, but a category three strike in between the two at Panama City and Destin. For an overview of the track history of Hurricane Eloise, see the Pensacola section in this chapter. The twelve- to sixteen-foot storm surge, a record at its time, was especially destructive between Fort Walton Beach and Panama City, and by comparison, Apalachicola escaped with only minor waterfront damage.

September 1, 1985: Hurricane Elena; category two strike at Apalachicola

Because of the bizarre looping track of Hurricane Elena, the storm generated hurricane strikes all the way from Cedar Key eastward to beyond Gulfport, Mississippi. Category three strikes occurred along the coast from Pensacola westward to the vicinity of Biloxi in Mississippi. At Apalachicola, Elena was a category two strike. See the Pensacola section of this chapter for an overview of this difficult-to-forecast hurricane track.

On September 1, when Elena was moving westward over the northern Gulf, the eye was only fifty miles south of Apalachicola for several hours. The town and adjacent coastal areas were again significantly damaged by the hurricane.

November 21, 1985: Hurricane Kate; category two strike at Apalachicola

Very late in the season, Hurricane Kate was classified as a tropical storm northeast of Puerto Rico on November 15. Four days later, Kate was a category one, and then two, hurricane over the Florida Straits north of Havana. Kate then intensified to category three over the Gulf, about 350 miles south of Apalachicola on November 20. Kate weakened as it swept northward toward the Florida Panhandle, coming on shore as a category one hurricane west of Port St. Joe the following day. Kate represented the fifth hurricane strike in twenty years at Apalachicola, beginning with Hurricane Alma in 1966.

The storm surge of eight feet or more was particularly destructive around Apalachicola, and beach erosion was especially severe at Cape San Blas and St. George Island. In addition, high winds were very destructive at Apalachicola, with the courthouse and municipal water tower sustaining much damage.

Along the northern Gulf Coast, hurricane strike frequencies increase sharply westward from Apalachicola. At Panama City Beach, for example, there have been fourteen hurricane strikes, one of which was a major hurricane. Farther west at Destin, for the same time period, there were thirteen strikes, but three were major hurricanes.

Pensacola and Pensacola Beach, Florida

The city of Pensacola is positioned on one of the few deep-water harbors, Pensacola Bay, around the shores of the Gulf. Pensacola is sheltered from the open waters of the Gulf by a chain of barrier islands, with Santa Rosa Island and Pensacola Beach across the bay from the city. The advantages of the harbor were discovered as early as 1540 by Spanish explorers, with permanent settlement dating back to 1698. Much of the city is positioned on bluffs above hurricane storm surges, but commercial and residential developments in low-lying areas adjacent to the harbor have been subject to destructive hurricane storm surges from time to time. Especially since World War II, Santa Rosa Island has been developed for beachfront recreation. The towns of Pensacola Beach and Navarre Beach—both characterized by mixtures of older, small beach cottages and motels, more recent townhouse and condominium complexes, and, most recently, high-rise condominiums on the beachfront—are extremely vulnerable to the onslaught of hurricane winds and storm surges (see fig. 6.5). Although the individual storm strikes vary from place to place, the strike record at Pensacola Beach is representative of the resort coast from Panama City Beach westward to Biloxi, Mississippi, a 250-mile strip of dense residential and commercial development in recent decades.

The frequency of strikes continues to increase toward the west. At Pensacola Beach there have been seventeen hurricane strikes, and five were classified as major hurricanes (see fig. 6.6).

September 27, 1906: storm 6; category two strike at Pensacola Beach

This hurricane was generally considered to be the worst storm in history as of that time at Pensacola. It was first identified as a tropical storm over the extreme western Caribbean off Nicaragua on September 19, becoming a hurricane over the Yucatan Channel on September 23, and then a

Fig. 6.5. Pensacola Beach, Florida.

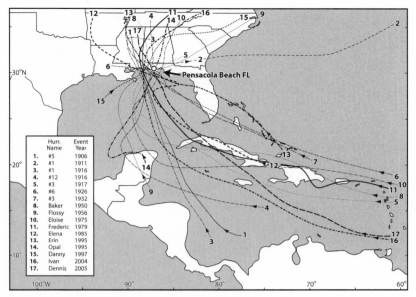

Fig. 6.6. Hurricanes having impact at Pensacola, Florida, 1901–2007.

major hurricane over the Central Gulf two days later, with landfall near Pascagoula, Mississippi, seventy-five miles west of Pensacola, on September 27. Nevertheless, a ten-foot storm surge destroyed most of the waterfront areas, more than 45 ships in the harbor were swept on shore, and the city recorded 32 deaths.

August 11, 1911: storm 2; category one strike at Pensacola Beach

This storm was initially identified as a tropical storm about one hundred miles southwest of Fort Myers on August 9. After intensifying to a category one hurricane about 150 miles southeast of Pensacola on August 11, the minimal hurricane came on shore the same day in the vicinity of Gulf Shores, Alabama, about thirty miles west of Pensacola. It was a small hurricane; nevertheless, there was considerable damage to property in the harbor and the city.

July 5, 1916: storm 2; category one strike at Pensacola Beach

In less than fifteen months, Pensacola was about to be battered by three hurricane strikes. The first of these storms followed a track much the same as that of the hurricane of 1906, originating over the western

Caribbean, reaching major hurricane status over the Central Gulf, 250 miles south of Pensacola on July 4, and making landfall the following day near Biloxi, Mississippi, about one hundred miles west of Pensacola. Despite the distance from landfall, Pensacola was again in the much more dangerous eastern sector of the hurricane, with sustained winds over one hundred miles per hour. However, the storm surge was less than in 1906, and there was less damage and almost no loss of life locally.

October 18, 1916: storm 14; category two strike at Pensacola Beach

Little more than three months later, Pensacola was again battered by a hurricane from the western Caribbean that made landfall at Pensacola Beach. Although the maximum sustained wind was recorded at 114 miles per hour, the much smaller hurricane and an excellent pre-storm forecast by the Weather Bureau resulted in no loss of life in Pensacola. There was only a three-foot storm surge and much less damage than the July hurricane.

September 29, 1917: storm 3; category three strike at Pensacola Beach

The third of the cluster of three hurricane strikes, this storm was first identified over the Atlantic just east of Guadeloupe on September 20. It then crossed most of the Caribbean, becoming a major hurricane south of Cuba on September 24. The storm then turned sharply northward, continuing as a category three on September 27, 275 miles southwest of Pensacola. It made landfall on September 29 in the vicinity of Pensacola again, as a category three storm. Wind speeds were again recorded just over one hundred miles per hour, and there was much more damage to waterfront areas than during the October storm of the previous year. Again, there was no loss of life in the Pensacola area.

September 21, 1926: storm 6, The Great Miami Hurricane; category three strike at Pensacola Beach

This hurricane evolved to tropical storm strength over the tropical Atlantic, midway between Africa and Florida, on September 11, and strengthened to a category four hurricane over the Bahamas on September 16. Without much prior warning to the public, this hurricane made its first landfall at Miami as a category four hurricane. The Florida "Gold Coast" from Palm Beach to Miami had been filling up with newcomers from the North who had no experience with hurricanes; most did not evacuate coastal locations and were caught outside as the eye passed directly over Miami. In terms of buildings and infrastructure, this storm was the most costly in the United States at the time, and the hurricane is given credit for ending the exploding real estate boom there.

The hurricane passed over southern Florida and out over the northeastern Gulf, where it made a second landfall on September 21 in the vicinity of Gulf Shores, Alabama. The hurricane center passed very close to Pensacola on its track to Alabama, and is counted as a category three strike at Pensacola. There were no recorded deaths due to the hurricane in Pensacola, but nearly all of the buildings around the harbor, as well as the vessels anchored there, were destroyed.

September 1, 1932: storm 3; category one strike at Pensacola Beach

Storm 3 developed initially as a tropical storm north of Haiti on August 26. After crossing southern Florida as a tropical storm, it intensified to a category one hurricane over the Gulf about one hundred miles southwest of St. Petersburg on August 30. The storm made landfall on September 1 near Gulf Shores as a category one hurricane, with moderate damage to waterfront buildings in Pensacola.

August 31, 1950: Hurricane Baker; category one strike at Pensacola Beach

By August 21, Hurricane Baker had intensified to a major hurricane northeast of Guadeloupe. Hurricane Baker progressed westward into areas with unfavorable environments for the maintenance of the storm, and, by August 26, Baker was only a tropical storm south of Cuba over the northwestern Caribbean. Two days later, however, over the Gulf just north of the Yucatan, Baker regenerated to a hurricane and by August 30 intensified to category two, only two hundred miles southwest of Pensacola. Hurricane Baker made landfall at Gulf Shores the following day as a category one hurricane. Baker resulted in severe damage around Pensacola and especially on the barrier islands from Pensacola eastward to beyond Destin.

September 24, 1956: Hurricane Flossy; category one strike at Pensacola Beach

Flossy originated as disturbed tropical weather over the eastern Pacific south of Guatemala. Tracking northward during passage over Guatemala, the disturbance became a tropical storm over the Gulf, just off the Yucatan, on September 22, became a minimal hurricane south of the

Mississippi River delta, made an initial landfall near Boothville, Louisiana, grazed New Orleans, and finally moved sharply to the northeast, making a second landfall as a category one hurricane at Fort Walton Beach, forty miles east of Pensacola. Flossy was the first hurricane to cause significant disruption to oil and gas extraction from the Gulf, but only minor waterfront damage was experienced around Pensacola. Heavy coastal rainfall was associated with Flossy, with a maximum of 16.3 inches at Gulf Shores, Alabama.

September 23, 1975: Hurricane Eloise; category one strike at Pensacola Beach

The barrier islands along the central Gulf Coast began to be developed intensively for recreation and beach vacations after World War II, and Hurricane Eloise was the first major hurricane strike there in about fifty years. Now there were thousands of residents and visitors at vulnerable locations on the barrier islands and waterfront locations around the bays and sounds. Eloise grew to hurricane strength over the central Gulf on September 22, intensified to major hurricane status only 140 miles south of Pensacola on September 23, and came ashore the same day as a category three hurricane near Destin, only fifty miles east of Pensacola. The high winds and twelve- to sixteen-foot storm surge from Fort Walton Beach to Panama City Beach were extremely destructive to almost all waterfront structures, but Pensacola, in the traditionally weaker western side of the storm, escaped with only minor damage and no loss of life.

September 13, 1979: Hurricane Frederic; category two strike at Pensacola Beach

Frederic was first identified as a tropical storm over the middle Atlantic on August 29, but it did not reach the Gulf as a category one hurricane until September 10. By two days later, Frederic had exploded to category four status only 225 miles south of Pensacola, with landfall as a major hurricane the following day at Dauphin Island, Alabama, about fifty miles west of Pensacola. More than half a million residents evacuated coastal areas from Louisiana eastward to Panama City Beach, the largest evacuation up to that time. The twelve-foot storm surge destroyed all the beachfront structures on Dauphin Island and eastward beyond Gulf Shores. The surge was also extremely destructive along the shores of the Florida Panhandle for about 150 miles, with more than five hundred homes in Pensacola and Pensacola Beach declared uninhabitable. There were very few lives lost because of the effectiveness of warnings and subsequent evacuations.

September 2, 1985: Hurricane Elena; category two strike at Pensacola Beach

Hurricane Elena threatened residents and Labor Day vacationers with evacuations all the way from Morgan City, Louisiana, to Sarasota, Florida, a coastline distance of more than 700 miles. About 1.5 million people evacuated, with residents and vacationers from Gulfport, Mississippi, eastward to Panama City Beach, including Pensacola, forced to evacuate two times in three days.

Elena developed as a tropical storm over the Florida Straits on August 29, moving northwestward so that two days later it was only one hundred miles southeast of Apalachicola. At category three, the storm threatened the entire Florida Panhandle with a major hurricane strike. At this point, Elena stopped forward movement and began drifting northeastward toward Cedar Key, prompting extension of evacuation orders southward as far as Sarasota and the lifting of evacuation orders along the northern Gulf Coast, including Pensacola and Pensacola Beach. Residents and vacationers returned. But much to the surprise of hurricane forecasters and residents, Elena executed a tight loop and abruptly resumed rather rapid motion to the northwest again, requiring second orders for evacuation along the northern Gulf Coast, including Pensacola and the beaches on the barrier islands. Elena swept along the coast from Destin westward to Gulfport, coming on shore near Biloxi as a category three storm. Despite storm surge damage to waterfront homes and buildings from near St. Petersburg all the way north and west to Gulfport, there were no deaths directly associated with the hurricane.

August 4 and October 4, 1995: Hurricanes Erin and Opal; categories one and three at Pensacola Beach

Sixteen years after Frederic, Pensacola was struck by two hurricanes two months apart in 1995. Erin was first identified as a tropical storm over the southern Bahamas on July 31. Three days later Erin made its first landfall near Vero Beach as a category one hurricane. After crossing the Florida Peninsula and the northeastern Gulf, Erin came on shore near Fort Walton Beach, forty miles east of Pensacola, as a minimal category two storm. The storm surge at Navarre Beach was six to seven feet, and about two-thirds of all buildings and other structures there were heavily damaged. To the west at Pensacola Beach, the surge was three to four feet, with, nevertheless, considerable damage to beachfront homes and condominiums.

The hurricane season of 1995 was extremely active, and Opal evolved into a tropical storm just off the northern coast of Yucatan on September 30. Two days later Opal had strengthened to a category one hurricane, still close to the Yucatan. Opal then began to sweep northward over the Gulf, intensifying to category four, almost a five, on October 4, only 250 miles southwest of Pensacola. At this time, however, Opal began to weaken and made landfall as a category three at Navarre Beach, essentially the same beach where Erin had made landfall two months earlier. Because Opal intensified so rapidly overnight, massive evacuations began in the morning, resulting in hours-long gridlock along the highways and roads leading northward from the coast. Some evacuees experienced the worst of hurricane conditions in their vehicles stranded along rural roads without shelters. At the coast, winds were weakened from the high winds offshore that same morning, but the storm surge reflected the earlier category four status and was as high as fifteen feet along the beaches of Santa Rosa and Okaloosa Island. All the buildings and other structures from Pensacola Beach eastward for about 120 miles were destroyed or heavily damaged, but again, there was no loss of life directly from the storm along the coast.

July 19, 1997: Hurricane Danny; category one at Pensacola Beach

Danny was the third hurricane strike at Pensacola in a little less than two years. Danny was first identified as a small tropical storm 150 miles south of Morgan City, Louisiana, on July 17. Danny drifted very slowly northeastward just south of the coast and intensified to a minimal hurricane over Boothville, Louisiana, the following day. On July 19, Danny, still a category one storm, made landfall in the vicinity of Gulf Shores, Alabama, after drifting over Mobile Bay for ten hours. Danny generated prodigious rainfalls, up to thirty-seven inches on Dauphin Island, but fortunately, because most of the very heavy rains fell close to the coast, flooding was not widespread. Hurricane Danny was also responsible for major damage to commercial fishing fleets at Grand Isle and Plaquemines Parish in southeastern Louisiana.

September 16, 2004: Hurricane Ivan; category three at Pensacola Beach

Hurricane Ivan was a major and devastating hurricane strike in the greater Pensacola area. Ivan is discussed as a memorable hurricane in chapter 5.

July 10, 2005: Hurricane Dennis; category three at Pensacola Beach

Less than one year later, another major hurricane, Hurricane Dennis, struck the greater Pensacola area. See chapter 5.

August 29, 2005: Hurricane Katrina; tropical storm at Pensacola Beach

Although Katrina was only a tropical storm strike at Pensacola, its massive storm surge further increased coastal erosion and did additional damage to shorefront properties in the greater Pensacola area and as far east as Panama City Beach. See chapter 2.

Dauphin Island, Alabama, located close to Mobile, has a strike record similar to that of Pensacola Beach, with fourteen hurricane strikes, five of them major. Gulfport, on the western Mississippi coast, is somewhat protected from the potential onslaught of hurricanes by the southeastward extension of the Mississippi River delta. Gulfport has experienced ten hurricane strikes, but only two, Camille in 1969 and Katrina in 2005, have been classified as major hurricane landfalls.

New Orleans, Louisiana

The geographical setting of New Orleans, between the Mississippi River to the south and Lake Pontchartrain to the north, makes New Orleans more storm prone for hurricane disasters than any other major American city. Most of the city and metropolitan area are below sea level. These areas are protected from river flooding and hurricane storm surges by a network of ring levees built by the U.S. Army Corps of Engineers and maintained by local levee boards. In addition, there is no natural drainage out of the city, and all of the storm runoff has to be pumped over the levees into Lake Pontchartrain and the Mississippi River. But unlike the other coastal towns and cities analyzed in this chapter, New Orleans is positioned about fifty miles north and west of the open coast, and tropical storms and hurricanes that threaten the metropolitan area usually weaken significantly before destructive hurricane winds reach the city. In addition, the coastal wetlands between the city and the open Gulf provide some resistance to storm surges, usually reducing the height of surges as they approach the city.

We identify and evaluate tropical storm and hurricane strikes for New Orleans back an additional fifty years to 1851 because of the historical role of New Orleans as the primary seaport and largest city on the Gulf Coast during most of the nineteenth century. Figures 6.7a–c show the tropical storm and hurricane strikes for 157 years between 1851 and 2007 at Morgan City, Boothville, and Gulfport. The figures compare three relatively open coast communities to the southwest, southeast, and east of New Orleans, respectively, with a southwest–northeast straight-line distance of about 150 miles between Morgan City and Gulfport. Tropical storm and hurricane strikes at these three coastal locations obviously represent potential storm threats to the city of New Orleans, and the figure also shows our estimates of tropical storm and hurricane strikes over the city directly.

Over the 157 years, there have been 96 different tropical storm and hurricane strikes at the three coastal sites, a little more frequent than once every other year on average, and each a potential threat to New Orleans. Of the 96 threats, 52 were tropical storms, 32 were category one or two hurricanes, and 12 were major hurricanes. In terms of our strike model for New Orleans, by contrast, there were 38 tropical storm strikes, only 15 category one or two hurricane strikes, and no major hurricane strikes. The low ratio of strikes to threats, and the strikes' lower intensities, are due to storms passing near but not over the city, and to the inland location of the city relative to the open coast. Hence one can appreciate the prevailing attitude of local residents before Katrina that storms would curve away and miss the city, that the storms would be much weaker over the city than forecasted and observed over adjacent Gulf waters, and that the levee and pumping complex would protect the city from significant flooding.

Prior to the beginning of our analysis starting with 1851, there is, nevertheless, a substantial history of devastating tropical storm and hurricane strikes at New Orleans, as reported by David Ludlum in *Early American Hurricanes, 1492–1870*. The earliest firsthand account of a strong hurricane is from September 23–24, 1722, when a hurricane coming on shore just to the west of New Orleans flattened and flooded all of the original temporary buildings erected at the newly established French capital of Louisiana. The Treaty of Paris in 1763 transferred colonial authority over Louisiana from the French to the Spanish, and the Spanish administration had to cope with three tropical storms or hurricanes in successive years, 1778, 1779, and 1780, which laid waste to the emerging Spanish colonial capital of Louisiana. Of the three, the great hurricane of August 1779 was the most severe and destructive in New Orleans, with

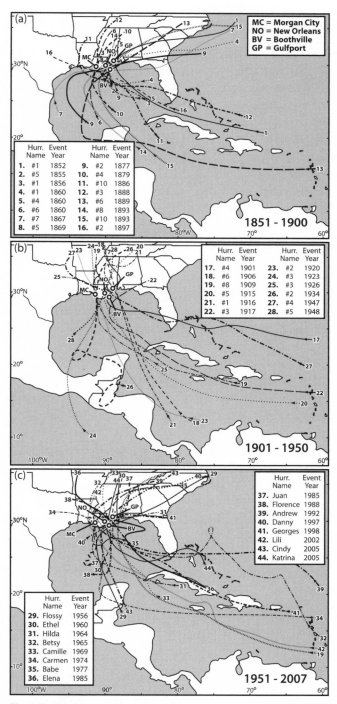

Fig. 6.7. Hurricanes either threatening or having impact at New Orleans, Louisiana: a) 1851–1900; b) 1901–1950; and c) 1951–2007.

most buildings and nearby agricultural crops destroyed and significant loss of life. The last hurricane of the eighteenth century occurred on August 31, 1794, when much of the Mississippi River delta below New Orleans was inundated by a widespread storm surge and assaulted by hurricane winds.

Again, as reported by David Ludlum, at least five hurricanes swept over southeastern Louisiana and the adjacent Mississippi Coast during the first half of the nineteenth century. These hurricanes certainly impacted New Orleans, which during the first half of the century had grown to become the largest city and port on the Gulf of Mexico and, indeed, one of the largest cities and ports of the expanding United States. Early during the War of 1812, an August 19 hurricane dispersed converging British naval vessels as the storm approached the Mississippi River delta, but at the same time the storm severely damaged many of the major buildings in New Orleans and damaged or destroyed at least fifty-three vessels on the river at New Orleans. Severe hurricanes again visited southeastern Louisiana and New Orleans in 1831 and most notably in 1837, when the "Racer Hurricane" severely ravaged the Gulf Coast, all the way from the mouth of the Rio Grande, near present-day Brownsville, through Galveston and New Orleans to Mobile and Pensacola. The Racer Hurricane was particularly destructive in New Orleans, with severe overflow flooding from Lake Pontchartrain and steamboat wrecks on the lake and river.

During the eight years between 1852 and 1860, there was a cluster of six hurricane threats, with four resulting in hurricane strikes at New Orleans (see fig. 6.8). Altogether, for the twenty-five years between 1851 and 1875, there were twelve tropical storm and hurricane threats; of these, four were tropical storms, five were category one or two hurricanes, and three were major hurricanes. In terms of strikes at New Orleans, however, there were three at tropical storm intensity and four category one or two hurricanes—a very active run of tropical storm and hurricanes for New Orleans.

September 16, 1855: storm 5; category two strike at New Orleans

This storm was recognized as a hurricane on September 15, 1855, about 250 miles south-southeast of New Orleans, only one day prior to a category three strike just to the east of New Orleans at Bay St. Louis, Mississippi, and a category two strike at New Orleans.

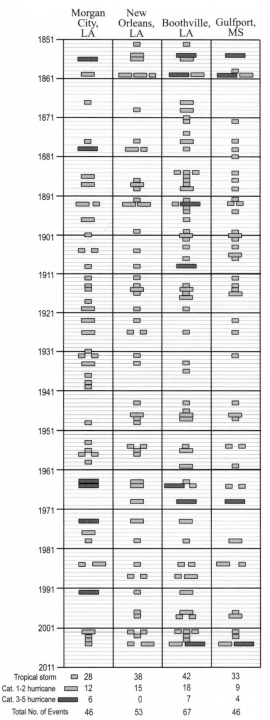

	Morgan City, LA	New Orleans, LA	Boothville, LA	Gulfport, MS
Tropical storm	28	38	42	33
Cat. 1-2 hurricane	12	15	18	9
Cat. 3-5 hurricane	6	0	7	4
Total No. of Events	46	53	67	46

Fig. 6.8. Timeline of tropical storm, hurricane (category 1–2), and severe hurricane (category 3–5) strikes at Morgan City, New Orleans, and Boothville, Louisiana, and Gulfport, Mississippi, 1851–2007.

August 10, 1856: storm 1, Isle Derniere Hurricane; category one strike at New Orleans

The most devastating and remembered of these four hurricanes is the "Last Island" or "Isle Derniere" hurricane of August 10, 1856. The barrier island is located about 70 miles southwest of New Orleans, and the storm surge suddenly swept over the entire barrier island destroying the resort "hotel" and the twenty cottages on the island. About 320 people were on the island when the hurricane struck. Many of the 180 survivors clung to wreckage of the ferry steamer *Star* that was swept onto the beach near the hotel. Legends of this storm were preserved in the romanticized novel *Chita: A Memory of Last Island,* by Lafcadio Hearn, published in 1889. Our model indicates a category one hurricane strike at New Orleans, but relative to the devastating events at Last Island, New Orleans residents probably thought about the storm as a near miss for the city itself.

August 12, 1860: storm 1; category one strike at New Orleans

The first of two hurricane strikes in 1860, this storm was at tropical storm strength 150 miles south of Apalachicola on August 7. Drifting westward ever so slowly, hurricane status was achieved on August 9, 200 miles south of Pensacola. One day later and still drifting westward, the hurricane intensified to category three, only 175 miles south of New Orleans. The first landfall, category three, was near Grand Isle on August 11, and the second landfall, still category three but with a track toward the northeast, was near Biloxi the following day. New Orleans, on the western side of the track, escaped with a category one strike.

September 15, 1860: storm 4; category one strike at New Orleans

This second hurricane of the season near New Orleans was identified first as a category two hurricane 150 miles west of Key West on September 11. After rather slow movement northwestward across the Gulf, the hurricane made its first landfall in lower Plaquemines Parish at Boothville on September 15 as a category two storm. Later the same day, the storm made a second landfall as a category one to the northeast, in the vicinity of Gulfport, and again New Orleans escaped with just a category one strike.

During the twenty-five years between 1876 and 1900, New Orleans was threatened by ten tropical storms, six category one or two hurricanes, and two major hurricanes—in total, eighteen storm threats in only twenty-five years (see fig. 6.8). Our strike model indicates six tropical storm and four category one strikes at New Orleans during these years.

September 1, 1879: storm 4; category one strike at New Orleans

The tropical storm was identified on August 28 over the central Gulf about 500 miles south-southeast of New Orleans. One day later it had intensified to a minimal hurricane, and on August 31, 275 miles southwest of New Orleans, to major hurricane status. The following day the category three hurricane was only 50 miles southwest of Morgan City, with landfall later on September 1 as a category three hurricane at Morgan City, 80 miles southwest of New Orleans, resulting in a category one strike at New Orleans.

August 19, 1888: storm 3; category one strike in New Orleans

On August 14 this storm reached tropical storm status east of the southern Bahamas. It intensified to major storm status only 150 miles southeast of Miami on August 16, with landfall at Miami later the same day. The hurricane exited Florida as a category one near Fort Myers. On its path toward the northwest, the hurricane eventually made landfall on August 19 near Grand Isle as a category two storm, resulting in a category one strike at New Orleans.

September 7, 1893: storm 8; category one strike in New Orleans

This storm formed over the western Caribbean east of Belize on September 4 and, after crossing the Yucatan, developed to hurricane status over the western Gulf of Mexico the next day. As the hurricane moved northeastward toward New Orleans, it intensified to category two status on September 6, which it maintained until just before landfall in the vicinity of Morgan City on September 7.

October 2, 1893: storm 10, Cheniere Caminada Hurricane; category one strike in New Orleans

Though a "near miss," the Cheniere Caminada Hurricane on October 2, 1893, was the most memorable hurricane of its time in New Orleans. It is described in chapter 5.

During the first twenty-five years of the twentieth century, there were ten tropical storm threats, seven category one or two hurricane threats, and just one major hurricane threat (see fig. 6.8). For the same twenty-five years in New Orleans, there were seven tropical storm strikes but just one category one hurricane strike.

September 29, 1915: storm 6; category one strike at New Orleans

This very deadly hurricane across southeastern Louisiana and coastal Mississippi is discussed in chapter 5.

During the twenty-five years from 1926 to 1950, there were eleven tropical storm threats, four category one or two hurricane threats, and no major hurricane threats (see fig. 6.8). Strikes at New Orleans amounted to seven tropical storms and one hurricane.

September 19, 1947: storm 4; category one strike at New Orleans

The September 1947 hurricane had its origins near the Cape Verde Islands in the eastern Atlantic on September 5, remaining at hurricane status for fourteen consecutive days as it crossed the Atlantic to make landfall at Fort Lauderdale, Florida, as a category four storm on September 17. It then emerged over the eastern Gulf, maintaining category two status over the Gulf south of New Orleans before weakening and coming on shore as a category one. This hurricane is especially remembered in Florida for maintaining category five status for two days east of Florida just prior to landfall. At New Orleans, nevertheless, a record storm surge of sixteen feet in Lake Pontchartrain flooded much of New Orleans, including Metairie and Moisant Airport (now Louis Armstrong New Orleans International Airport), with twelve people drowning in New Orleans and southeastern Louisiana.

The second half of the twentieth century and the first five years of the twenty-first were marked with two extraordinary clusters of hurricane threats and strikes. The first cluster, occurring between 1964 and 1974, is remembered for the threats of four major hurricanes—Hilda (1964), Betsy (1965), Camille (1969), and Carmen (1974)—each of which resulted in category one hurricane strikes in New Orleans. The second cluster, occurring between 1992 and 2005, included seven major hurricane threats—Andrew (1992), Georges (1998), Isidore and Lili (2002), Ivan (2004), and Katrina and Rita (2005)—with only Katrina resulting in a hurricane strike for the city.

For the twenty-five year span 1951–1975, there were six tropical storm threats, two category one or two hurricane threats, and four major hurricane threats. In the city of New Orleans, by contrast, there were four tropical storm strikes and four category one or two hurricane strikes (see fig. 6.8).

September 23, 1956: Hurricane Flossy; category one strike at Boothville on the Mississippi River below New Orleans

Coming from the southwestern Gulf, Flossy made initial landfall as a category one hurricane in lower Plaquemines Parish with 16.7 inches of tropical rain at Golden Meadow, a storm surge that totally inundated Grand Isle, and much beach erosion that resulted in one of the first beach nourishment programs along the shores of the Gulf. Although hurricane-force winds were not recorded in New Orleans, a seawall was overtopped, resulting in the flooding of about 2.5 square miles of the city. Flossy was the first Gulf hurricane to significantly affect oil and gas production from the Gulf, with production shut down for several days and some minor damage to platforms.

October 3, 1964: Hurricane Hilda; category one strike in New Orleans

Hilda, the first major hurricane of the 1964–1974 cluster, became a tropical storm over the Yucatan Channel on September 29, 1964. Hilda exploded rapidly to a category four hurricane two days later over the central Gulf, about 400 miles south of New Orleans. As Hilda approached landfall on October 3, 75 miles south of Morgan City and 150 miles southwest of New Orleans, the storm weakened, resulting in a category three strike at Morgan City and a category one at New Orleans. As the storm approached Louisiana, Hilda crossed over the newly developing offshore oil and gas fields, forcing evacuation of more than two thousand offshore workers and causing the complete destruction of thirteen oil and gas platforms. New Orleans itself escaped with mostly minor damage, but several tornadoes were very destructive. One at Larose, 30 miles southwest of New Orleans, killed twenty-four people. A water tower at Erath, 20 miles south of Lafayette, collapsed onto the roof of the city hall, killing eight more people.

September 10, 1965: Hurricane Betsy; category one strike in New Orleans

This was another very deadly hurricane in New Orleans after a category three strike at Grand Isle, Louisiana. See chapter 5.

August 17, 1969: Hurricane Camille; category one strike in New Orleans

A record-tying category five at first landfall in Plaquemines Parish, Louisiana, and again at second landfall at Bay St. Louis, Mississippi, Hurricane Camille brushed by New Orleans as a category one strike. Damage to the city was surprisingly light because of its position on the weaker western side of this very intense but small hurricane. See chapter 5.

September 7, 1974: Hurricane Carmen; category one strike in New Orleans

Hurricane Carmen was a major hurricane that threatened New Orleans, coming very close before turning sharply to the west at the last moment. New Orleans experienced only some heavy flooding rains and minor damage. Carmen was first identified as a tropical storm over the Caribbean, south of Puerto Rico, on August 30. Two days later Carmen threatened Belize as a category four hurricane, but then weakened and turned a bit to the north, with landfall in the southern Yucatan Peninsula of Mexico, north of the Belize border. Carmen emerged over the Bay of Campeche as a tropical storm on September 3. Moving slowly northward, Carmen threatened New Orleans on September 7 as a category four hurricane, with the center only 175 miles south of the city. The hurricane made landfall the following day near Morgan City as a category three storm but turned sharply to the west, leaving New Orleans with a weak category one strike, a six-foot storm surge, heavy rains, and limited damage.

Hurricane Carmen is remembered in Louisiana for severe damage to the sugarcane crops, shrimp harvests, and offshore oil and gas platforms. The hurricane is also memorialized in the popular 1994 film *Forrest Gump,* in which it is credited with destroying all but one of the shrimping boats in much of southeastern Louisiana.

During the 25-year period of 1976–2000, there were only four tropical storm threats, but six category one and two hurricanes threats, and one major hurricane threat (fig. 6.8). In the city of New Orleans there were seven tropical storm strikes and no hurricane strikes.

August 25, 1992: Hurricane Andrew; tropical storm strike in New Orleans

As a category four hurricane over the Gulf, Andrew approached and threatened New Orleans on August 25, 1992. The storm was located only 175 miles south-southeast of New Orleans. Fortunately for the city, Andrew swept farther to the west and made landfall west of Morgan City as a category three storm, giving New Orleans a tropical storm strike with minimal damage. Andrew is discussed as a memorable hurricane in chapter 5.

September 27, 1998: Hurricane Georges; tropical storm strike in New Orleans

Hurricane Georges intensified to a category five storm on September 20, over the western Atlantic about 300 miles east of Guadeloupe. After crossing the Florida Peninsula, Georges maintained category two status until the center was only 75 miles east-southeast of New Orleans, with a tropical storm strike for the city September 27–28. The storm surge was very damaging to restaurants and "fishing camps" in New Orleans East along the southern shore of Lake Pontchartrain.

During the seven years between 2001 and 2007, there were seven tropical storm threats, two category one and two hurricane threats, and one major hurricane threat (see fig. 6.8). The impacts in New Orleans included four tropical storm strikes and one category two hurricane strike.

September 26, 2002: Hurricane Isidore; tropical storm strike in New Orleans

On September 23, Hurricane Isidore was a category three hurricane off the northern coast of the Yucatan Peninsula, 600 miles south of New Orleans. Isidore, however, weakened to tropical storm status as it moved northward across the Gulf, making landfall just east of New Orleans. Isidore is remembered for local rainfall of over twenty inches and much local flooding (see chapter 7).

October 3, 2002: Hurricane Lili; tropical depression strike in New Orleans

One week later Hurricane Lili became another serious threat to New Orleans. Lili was a category four storm on October 2, only 175 miles southwest of the city. As Lili approached the Louisiana coast the following day, the storm weakened to category three, making landfall to the west of New Orleans near Morgan City as a category two. In New Orleans there were only minimum gale-force winds and three to seven inches of rain.

September 16, 2004: Hurricane Ivan; tropical depression strike in New Orleans

Hurricane Ivan was a very dangerous threat to New Orleans, with the center of the category three storm only ninety miles south of the city on

September 15. However, Ivan curved toward the northeast, making landfall in Alabama as a major hurricane, with New Orleans escaping again with an almost dry but windy day after massive evacuation efforts between the city and Baton Rouge. Ivan is discussed as a memorable hurricane in chapter 5.

August 29, 2005: Hurricane Katrina; category two strike in New Orleans

New Orleans was devastated by Hurricane Katrina, a category five storm less than one day before landfall, and a category two storm over the city. See chapter 2.

September 25, 2005: Hurricane Rita; tropical depression strike in New Orleans

Less than one month later, New Orleans was threatened again by category four and five Rita, which bypassed New Orleans on its track to landfall near the Louisiana-Texas border. Rita was close enough to New Orleans, however, for temporary levee repairs to be breached and for the reflooding of some of the city. Rita is discussed as a memorable hurricane in chapter 5.

Out very close to the open Gulf, Boothville, perched on the low natural levee of the Mississippi River Delta, has struggled for survival during sixteen hurricane strikes since 1900, of which four were major. A little to the west at Morgan City, there have been eight category one and two hurricane strikes together with four major hurricane strikes, for a total of twelve hurricane strikes. At Cameron, Louisiana, near the Texas border, the total number of strikes is only seven, two of which, Audrey in 1957, and Rita in 2005, were devastating major hurricanes.

Galveston, Texas

The city of Galveston is located on a barrier island of the same name and is separated from the mainland by Galveston Bay and West Bay. Prior to the deadly hurricane of 1900, Galveston was the leading American port on the western Gulf, the largest city in Texas, and a cosmopolitan city with commercial and trade connections worldwide. After the catastrophic hurricane of 1900, there have been thirteen hurricane strikes at Galveston, four of which are classified as major hurricane strikes. Figure 6.9 shows the tracks of the hurricane strikes at Galveston.

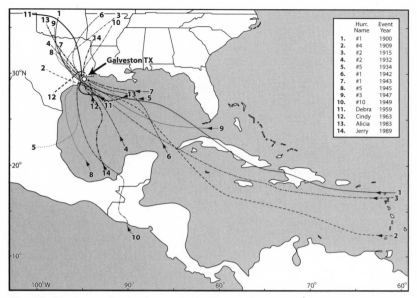

	Hurr. Name	Event Year
1.	#1	1900
2.	#4	1909
3.	#2	1915
4.	#2	1932
5.	#5	1934
6.	#1	1942
7.	#1	1943
8.	#5	1945
9.	#3	1947
10.	#10	1949
11.	Debra	1959
12.	Cindy	1963
13.	Alicia	1983
14.	Jerry	1989

Fig. 6.9. Hurricanes having impact at Galveston, Texas, 1901–2007.

September 8, 1900: storm 1; category three strike at Galveston

Chapter 1 focuses on this catastrophic hurricane strike that remains to this day the deadliest natural disaster in American history.

July 21, 1909: storm 4, the Valesco Hurricane; category two strike at Galveston

The Valesco Hurricane was first identified as a tropical storm over the western Caribbean on July 16, 1909. Two days later, the storm reached hurricane intensity near the western tip of Cuba, about 850 miles southeast of Galveston. By July 21, the hurricane had strengthened to category three, only 50 miles southeast of Galveston. It made landfall the same day at Valesco, now a neighborhood of Freeport, only 25 miles southwest of Galveston. Half of the small town of Valesco was destroyed, and forty-one people lost their lives in southeastern Texas. The storm's ten-foot storm surge at Galveston was the first successful test of the seawall built after the devastating hurricane of 1900.

August 17, 1915: storm 2; category three strike at Galveston

This hurricane was the first of two major hurricane strikes along the northern Gulf Coast during the 1915 season. The second strike was in southeastern Louisiana on September 29 and is discussed as a memorable hurricane in chapter 5. This hurricane was the second major test of the seawall and the elevation of the city following the Galveston Hurricane of 1900.

This Cape Verde Hurricane was first identified as a tropical storm off the west coast of Africa on August 5 and attained hurricane force northeast of Barbados on August 9. The hurricane swept over the northern Caribbean, interacting with Puerto Rico, Hispaniola, Jamaica, and Cuba, attained major hurricane status south of Cuba and north of Jamaica, and reached category four status over the central Gulf, 500 miles southeast of Galveston, on August 15. The hurricane made landfall as a category three storm about 25 miles southwest of Galveston on August 17.

The 16-foot surge damaged the seawall, which nevertheless held fast, with no more than five to six feet of flooding in the most low lying areas of the city. The hurricane, however, did remove most of the beach sand in front of the seawall, and the beaches have never recovered to this day. Only eleven lives were lost in Galveston, but between fifty and one hundred people drowned or remained missing from the unprotected western areas of Galveston Island. The causeway and water main from the mainland were heavily damaged, resulting in severe water shortages and delays in bringing help and supplies to the city.

August 13, 1932: storm 2; category four strike at Galveston

This small and short-lived hurricane was identified as a tropical storm north of Progreso in the Yucatan, about 700 miles southeast of Galveston, on August 11. During its track toward the Galveston area early on August 13, the storm intensified from a tropical storm to a category four hurricane in a single day, only 30 miles southeast of Galveston. The storm made landfall as a category four hurricane about 25 miles southwest of Galveston in the vicinity of Freeport. The intense but compact hurricane strike resulted in a 30- to 40-mile swath of coastal destruction between Galveston and Freeport, drowning forty people and flattening oil industry communities and facilities. However, damages in the city of Galveston, protected by the seawall, were minimal.

August 27, 1934: storm 5; category one strike at Galveston

The track of this minimal hurricane was bizarre. As a developing tropical storm, it was centered 450 miles southeast of Galveston on August 26. The following day, with the center only 25 miles southeast of Galveston, the storm briefly attained hurricane strength. It then drifted slowly toward the southwest along the southern Texas and northern Mexican coasts, first as a minimal hurricane and later as a weaker tropical storm, making final landfall near Tampico. After departing from the vicinity of Galveston, this track was the opposite of those of most of the tropical storms and hurricanes over the western Gulf. There was no significant damage in Galveston.

August 21, 1942: storm 1; category one strike at Galveston

This storm developed as a tropical storm over the Yucatan Channel on August 17, further developing into a category one hurricane over the Central Gulf, 300 miles southeast of Galveston two days later. On August 21, the minimal hurricane made landfall at Crystal Beach on the Bolivar Peninsula, 12 miles northeast of Galveston. The storm surge at Crystal Beach was seven feet, but damage in Galveston was minor. This storm was the first of a cluster of five hurricanes to strike Galveston in the eight years between 1942 and 1949.

July 27, 1943: storm 1; category one strike at Galveston

This was the so-called surprise hurricane because of the complete blackout of weather forecasts and news due to near-shore German submarine activity along the Texas and Louisiana coasts during World War II. This minimal hurricane made landfall at Bolivar Point, just northeast of Galveston, without any warnings to the local population. Postwar analyses indicated that the storm developed as a hurricane on July 25 southeast of the Mississippi River delta, and then proceeded on a northwesterly track over two days to reach the Galveston area. The northwesterly winds in Galveston generated a surge of only three to six feet that washed into the city from Galveston Bay rather than from the Gulf. Most likely because of the lack of advance warnings, nineteen people died from storm-related causes in southeastern Texas.

August 28, 1945: storm 5; category two strike in Galveston

First spotted as a tropical storm over the Bay of Campeche, 750 miles south of Galveston on August 24, this storm developed rapidly off the southeast Texas coast to a category four hurricane and made an initial strike at Port Aransas, on Mustang Island east of Corpus Christi, by August 27. The hurricane continued its northward path parallel with the coast and made landfall at Matagorda, 80 miles southwest of Galveston.

A fifteen-foot storm surge at Port Lavaca and nearly thirty inches of storm rainfall generated significant coastal damage and losses along the central Texas coast and caused three deaths at Port O'Connor. Galveston, behind its seawall, again escaped significant damage.

August 24, 1947: storm 3; category one strike at Galveston

This slow-moving storm developed into a tropical storm over the Florida Straits about 1,000 miles southeast of Galveston on August 18. Four days later, on August 22, the storm reached hurricane status about 400 miles southeast of Galveston. On August 24 the hurricane passed directly over Galveston Bay, with a storm surge of only three feet and only one storm-related death.

October 4, 1949: storm 10; category three strike at Galveston

This storm was first recognized as a tropical storm over the eastern Pacific, off the coast of El Salvador, on September 27. The circulation remained together as it crossed Central America and strengthened to a category one hurricane over the Bay of Campeche, 550 miles south of Galveston, on October 2. Moving rapidly northward, the hurricane attained major hurricane status on October 3 and reached category four just before landfall in the vicinity of Freeport, about 50 miles southwest of Galveston. The surge at Freeport reached eleven feet, but apparently because of good advance forecasts and a coastal population well acquainted with the cluster of storm strikes during the decade, only three lives were lost at Freeport. Again, probably due to the protection offered by the seawall, damage in Galveston was minor.

July 25, 1959: Hurricane Debra; category one strike at Galveston

Debra was another tropical event that brewed up quickly not far off the Texas coast. Debra was classified a tropical storm on July 23, only 175 miles southeast of Galveston. The following day, Debra intensified to a minimal hurricane 75 miles south-southwest of Galveston, and on July 25 the minimal hurricane made landfall along the western end of Galveston Island, with no significant damage and no fatalities.

September 11, 1961: Hurricane Carla; tropical storm strike at Galveston

Hurricane Carla is one of the most well known hurricanes along the Texas coast, with landfall as a category four storm near Port O'Connor, about 115 miles southwest of Galveston. Although only a tropical storm strike at Galveston, Carla was responsible for significant damage in the Galveston area and along the coast to the southwest. This storm is credited with the start of Dan Rather's fascination with hurricanes early in his career as a TV news personality.

September 17, 1963: Hurricane Cindy; category one strike at Galveston

Cindy was another short-lived storm off the southeastern Texas coast. Cindy developed into a tropical storm and a minimal hurricane on September 16, less than 200 miles south-southeast of Galveston. Cindy made landfall as a category one hurricane on High Island, about 35 miles northeast of Galveston. There was a four-foot storm surge at Galveston, resulting in limited structural damage but three drownings associated with 15- to 23-inch storm rainfalls and flooding.

August 17, 1983: Hurricane Alicia; category three strike at Galveston

Alicia was the first major hurricane strike after the cluster of strikes between 1942 and 1949. Another Galveston hurricane with a very short history, Alicia was identified as a tropical storm on August 15, 300 miles southeast of Galveston. The following day Alicia reached hurricane intensity 175 miles southeast of the city, and on August 17 Alicia exploded to major hurricane status with the eye only about 25 miles south of Galveston. Alicia came ashore near the western end of Galveston Island as a category three hurricane.

Alicia was a small but intense hurricane, with a twelve-foot storm surge on the Galveston Bay waterfront and an eight-foot surge later on the Gulf-side waterfront. Prior to the arrival of Alicia, the Galveston mayor had ordered evacuations from only the most low lying areas of the city, and only about 10 percent of the residents left ahead of the storm. After Alicia began to intensify offshore, evacuation became impossible because the bridges to the mainland were impassable.

Storm damage in the city was not severe with the exception of residences and commercial establishments immediately behind the seawall, due to destructive winds. But the residents of the western half of the island, much closer to landfall, and without the protective seawall, lost almost everything to the onslaught of wind and water. Twenty-one deaths were attributed to the hurricane. Meanwhile, inland at Houston, loose gravel on the rooftops of recently erected skyscrapers became tiny projectiles that shattered thousands of glass windows, making pedestrian move-ment on the streets below dangerous from the mix of airborne gravel and glass shards. The building code was belatedly corrected by the Houston city council.

October 15, 1989: Hurricane Jerry; category one strike at Galveston

Jerry, another minimal hurricane, became a tropical storm on October 12 over the Bay of Campeche, 700 miles southeast of Galveston. It was not until October 15, the very day of the hurricane strike, that Jerry achieved hurricane status, with the center only 90 miles south of the city. Despite landfall at Galveston, there was minimal damage to the city. Three per-sons were killed when their vehicle crashed off the seawall during the hurricane. Jerry was also responsible for the erosion of a twenty-mile sec-tion of Texas highway 87 between High Island and Sabine Pass. As of 2008, there are no firm plans to rebuild this most direct coastal link be-tween Galveston and Port Arthur, adjacent to the Louisiana border.

Along the coast southwest of Galveston, hurricane strike frequencies de-crease. Port O'Connor is located 115 miles southwest of Galveston, and the number of hurricane strikes decline from thirteen at Galveston to nine at Port O'Connor. Similarly, major hurricane strikes decrease from four at Galveston to two at Port O'Connor.

Port Aransas (Corpus Christi), Texas

Port Aransas, on a barrier island known as Mustang Island, is adjacent to Aransas Pass, which connects Corpus Christi Bay with the Gulf of Mex-ico. A deepwater channel is maintained so that ocean-going vessels have access to the large bay and the city of Corpus Christi, more than fifteen miles across the bay from Port Aransas. In Corpus Christi, Ocean Drive borders the crescent-shaped waterfront for miles, with spectacular mod-ern skyscrapers downtown and mansions and villas surrounded by sub-tropical landscapes on the land side of the boulevard. Port Aransas, on the barrier island, is very vulnerable to the winds and storm surges of hur-ricanes, whereas Corpus Christi is somewhat more sheltered. Some sec-tions of Ocean Drive are on bluffs forty feet above sea level and thus be-yond the potential domain of storm surges. Figure 6.10 shows the tracks of hurricanes identified as strikes at Port Aransas.

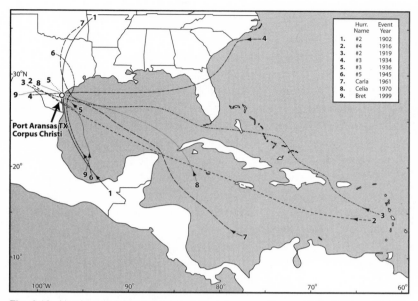

Hurr. Name	Event Year
1. #2	1902
2. #4	1916
3. #2	1919
4. #3	1934
5. #3	1936
6. #5	1945
7. Carla	1961
8. Celia	1970
9. Bret	1999

Fig. 6.10. Hurricanes having impact at Port Aransas, Texas, 1901–2007.

June 26, 1902: storm 2; category one strike at Port Aransas

This short-lived hurricane first reached tropical storm status over the Bay of Campeche, 550 miles south of Port Aransas, on June 24. Off the northern Mexican coast one day later, the storm strengthened to a minimal hurricane. The storm made landfall on Mustang Island as a minimal hurricane on June 26, with little to no significant damage of record.

August 18, 1916: storm 4; category one strike at Port Aransas

First identified as a hurricane north of Barbados on August 12, 1916, this storm gradually intensified during its westerly track, attaining major hurricane status 100 miles south of the Isle of Youth on August 16. Major hurricane status was maintained until hours before landfall, when the center was only 125 miles southeast of Port Aransas. Weakening as it approached the coast, the hurricane made a category two landfall on Padre Island about 60 miles south of Port Aransas on August 18. The hurricane wrecked some properties in both Port Aransas and Corpus Christi. Five people died in Corpus Christi and two more on a capsized fishing boat offshore.

September 14, 1919: storm 2; category three strike at Port Aransas

This hurricane battered Key West four days before landfall in southern Texas, and the description of its track to the Florida Keys can be found in the Key West section of this chapter. On September 13, one day before landfall on Padre Island, the category four hurricane was over the west-central Gulf, about 400 miles east-southeast of Port Aransas. At this time the Weather Bureau had "lost" the position and track of the hurricane, because the bureau depended entirely on reports from ships at sea, and there were almost no ships in the western Gulf. The bureau expected the hurricane to curve northward and strike Louisiana. One day before the strike in the Corpus Christi area, there were still no warnings of the rapid approach of this very dangerous hurricane, still a category three, less than 250 miles southeast of the city.

The hurricane made its second landfall in the United States on Padre Island less than 50 miles south of Corpus Christi. The storm surge at Port Aransas was sixteen feet, sweeping over everything on Mustang Island and breeching oil storage tanks at Port Aransas, causing a plume of heavy oil to cover everything in and around the shores of Corpus Christi Bay. The surge destroyed most buildings in downtown Corpus Christi, where debris piles were fifteen to twenty feet high. The official death toll at Port Aransas and in Corpus Christi was a little less than three hundred, but local residents believed the true loss of life was between six hundred and one thousand people. The disaster was downplayed, they believed, because the city was in the beginning phases of massive industrial and residential development and public officials did not wish to spread fears of the potential for future hurricane disasters.

July 25, 1934: storm 3; category one strike at Port Aransas

The track of this storm was bizarre, in that it was first identified as a developing tropical storm just south of Cape Hatteras on July 21. The tropical storm then proceeded southwestward and made an initial landfall near St. Augustine, Florida, crossed the peninsula, and moved over the Gulf at Cedar Key on July 24. Continuing on this westward track, the storm reached minimum hurricane status on July 25 about 150 miles east of Port Aransas, coming on shore there as a category one hurricane later the same day. In Texas eleven deaths were attributed to the storm, almost all in association with tornadoes.

June 27, 1936: storm 3; category one strike at Port Aransas

This storm was extremely short lived, being first identified as a tropical storm only 150 miles southeast of Port Aransas on June 26, 1936. Just before landfall at Port Aransas the following day, the storm intensified to a minimum hurricane. There were no deaths, and damage was restricted mostly to an oil refinery.

August 26, 1945: storm 5; category two strike at Port Aransas

The track history of this hurricane was discussed in the Galveston section of this chapter, with this hurricane a category two strike at Galveston. On August 26, the hurricane bypassed Port Aransas and Corpus Christi with the center of the category four hurricane offshore only forty miles east of Port Aransas. Despite the close call, there was minimal damage in Port Aransas and Mustang Island, with the cities sheltered in the weaker western side of the hurricane circulation.

September 11, 1961: Hurricane Carla; category two strike at Port Aransas

This infamous Texas hurricane evolved very quickly as a tropical storm and category one hurricane over the northwestern Caribbean on September 5 and 6, 1961, becoming a major hurricane the following day over the eastern Gulf. On September 10, one day before landfall, Carla attained the dreaded category five status, only 150 miles southeast of Port Aransas. The next day the center of Carla tracked northwestward only fifty miles off the coast at Port Aransas before making landfall as a category four hurricane in the vicinity of Port O'Conner, 75 miles northeast of Port Aransas.

For its time, Carla was the largest hurricane that had been observed over the Atlantic Basin and the strongest to strike Texas. There was significant coastal damage for the entire Texas coast from High Island to South Padre Island. The storm surge was as high as twenty-two feet around the head of bays, and at Corpus Christi there was significant wind damage. A gust of 170 miles per hour was observed at Port Lavaca, very close to landfall. Carla also spawned twenty-six tornadoes, including a deadly one at Galveston. Overall there were thirty-one storm-related deaths in Texas, undoubtedly kept down because of advance and persistent warnings.

August 3, 1970: Hurricane Celia; category two strike at Port Aransas

Celia developed as a tropical storm over the southeastern Gulf, 850 miles southeast of Port Aransas, on August 1, 1970. The very same day, Celia exploded to a category three hurricane. On its track to the Texas coast over the next two days, Celia weakened to a minimal hurricane and then intensified again to major hurricane status. Celia came within 60 miles of Port Aransas before landfall on August 3, just north of Port Aransas.

Celia was a very unusual hurricane at landfall over Corpus Christi, with extremely high, damaging bursts of wind up to 161 miles per hour, briefly, over the left-hand side of the hurricane circulation—normally the less dangerous and destructive side. In Corpus Christi, wind damage was extremely severe; one person died and nearly five hundred were injured. The storm surge at Port Aransas was only nine feet, less than would be expected for a hurricane of this intensity.

August 22, 1999: Hurricane Bret; category one strike at Port Aransas

Hurricane Bret first developed as a tropical storm over the Bay of Campeche, 600 miles south-southeast of Port Aransas, on August 19, 1999. On the way to Texas, Bret intensified rapidly to category four status 250 miles southeast of Port Aransas on August 21. At landfall the following day on South Padre Island, about 75 miles south of Port Aransas, Bret is credited with a category three strike. The landfall was in an area with the fewest residents of any coastal area in the forty-eight contiguous states of the United States. A six-foot storm surge resulted in significant beach erosion, and heavy rains destroyed irrigated crops inland. There was no loss of life associated with Hurricane Bret.

Padre Island, a very long barrier beach and island, extends southward more than 110 miles from Mustang Island and Corpus Christi to near Port Isabel. The barrier island is primarily taken up by Padre Island National Seashore, with resort communities at the northern and southern ends. The hurricane strike record at the southern end of South Padre Island is similar to that at Port Aransas, with the same number of strikes, nine. However, two of the nine strikes at South Padre Island were major strikes—Hurricane Beulah in 1967 and Hurricane Allen in 1980—as compared with none at Port Aransas. Hurricanes Beulah and Allen are discussed as memorable hurricanes in chapter 5.

The 300-mile coast of the Mexican state of Tamaulipas is sparsely inhabited except at Matamoros, upriver from the mouth of the Rio Grande, and at the southern end of the state, where the city of Tampico is located. Hurricane strike data for two small coastal settlements between Matamoros and Tampico show six strikes, including two major hurricane strikes at Boca Madre and seven strikes, of which three were major, at Lago Morales. For centuries along this coast, the loss of life and massive destruction during hurricane strikes have been caused not so much by winds and storm surges on the immediate coast, but rather by the torrential rainfalls and resultant massive flooding associated with moist air from the Gulf forced up and over the high and rugged Sierra Madre Oriental. In times of hurricane strikes, these conditions have overwhelmed coastal villages and cities located close to where the rivers approach the Gulf.

Tampico, Tamaulipas State, Mexico

Tampico is a thriving commercial and port city with a metropolitan population approaching one million people. It is the center of a rich petroleum-producing region, and exports oil, minerals, and agricultural products. The strike history for Tampico includes eleven hurricanes, two

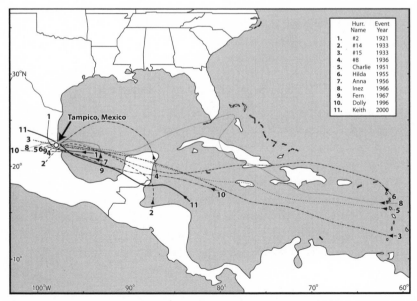

Hurr. Name	Event Year
1. #2	1921
2. #14	1933
3. #15	1933
4. #8	1936
5. Charlie	1951
6. Hilda	1955
7. Anna	1956
8. Inez	1966
9. Fern	1967
10. Dolly	1996
11. Keith	2000

Fig. 6.11. Hurricanes having impact at Tampico, Mexico, 1901–2007.

of them major. The tracks of these eleven hurricanes are shown in figure 6.11.

September 7, 1921: storm 2; category one strike at Tampico

This short-lived storm developed initially as a tropical storm on September 6 over the extreme western Gulf only 275 miles southeast of Tampico. Later the same day the storm attained minimal hurricane status, and it made landfall the next day about 25 miles south of Tampico. Although heavy flooding rainfalls did occur around Tampico, there was limited damage and no loss of life there. However, remnants of this hurricane interacted with a cold front over central Texas several days later, with a United States record rainfall of thirty-six inches in only eighteen hours at Thrall, between San Antonio, Austin, and Temple, resulting in the drowning of 215 people.

September 15, 1933: storm 14; category one strike at Tampico

On September 10, 1933, a tropical storm was identified over the northwestern Caribbean and the Gulf of Honduras. The storm drifted slowly toward the north and west, and then crossed the Yucatan Peninsula, where it was a category two strike at Cozumel. It then maintained minimal hurricane status over the Bay of Campeche. The minimal hurricane came on shore the following day as a strike only twenty miles north of Tampico. This hurricane resulted in heavy damage in Tampico and Tuxpan, with sixty-seven people drowned and thousands homeless.

September 24, 1933: storm 15; category one strike at Tampico

Only nine days after the previous hurricane, this second hurricane strike of the season made landfall directly at Tampico. This Cape Verde storm first evolved as a tropical storm over the far Atlantic on September 16, finally attaining hurricane status over the southern Caribbean 100 miles southwest of Kingston, Jamaica, on September 20. Four days later the hurricane reached Tampico as a category one strike.

Floodwaters—ten to fifteen feet deep—due to a combination of storm surge and runoff from rainfall covered most of the city. Almost all of the buildings were heavily damaged or destroyed, and casualties, both dead and injured, amounted to more than five thousand.

August 19, 1936: storm 8; category one strike at Tampico

On August 15 a tropical storm was identified over the Yucatan Channel about 800 miles east of Tampico. The next day a hurricane-strength storm was reported over the central Gulf about 600 miles east-northeast of Tampico. Landfall as a category one strike, again right at Tampico, was on August 19. This was the third hurricane strike in three years. The storm produced heavy rainfall and flooding, but no local deaths and little damage.

August 22, 1951: Hurricane Charlie; category three strike at Tampico

This storm was first identified as a tropical storm midway between Africa and the Windward Islands on August 14, 1951. By August 19, the hurricane achieved major status over the western Caribbean, 175 miles southwest of the Isle of Youth, and category four status the following day, then weakened dramatically before landfall on the Yucatan. On August 20, Charlie made landfall on the northeastern corner of the Yucatan Peninsula, weakening to category two as it crossed the peninsula and strengthening again over the Bay of Campeche, only 80 miles east of Tampico on August 22. Charlie became a category three hurricane strike directly over Tampico later the same day. At Tampico the hurricane generated heavy rains and river flooding, and more than one hundred people drowned.

September 19, 1955: Hurricane Hilda; category one strike at Tampico

This Hilda, not to be confused with Hurricane Hilda along the northern Gulf Coast in 1964, developed as a tropical storm over the Atlantic northeast of Puerto Rico on September 11, 1955. By September 15, it was a category three storm over the northwestern Caribbean Sea east of the Yucatan. After a brief bout with the Yucatan, Hilda regained major hurricane status, making landfall on September 19 as a category one strike at Tampico.

July 26, 1956: Hurricane Anna; category one strike at Tampico

Anna, a one-day hurricane, developed as a tropical storm on July 25, 1956, over the Bay of Campeche, only 325 miles southeast of Tampico. Anna reached minimal hurricane status the next day, with landfall as a minimal hurricane ten miles south of Tampico, less than one year after Hilda.

October 10, 1966: Hurricane Inez; category two strike at Tampico

Inez survived two weeks over the Atlantic, Caribbean, and Gulf as a hurricane, with a number of destructive landfalls. Inez appeared first as a tropical storm halfway across the Atlantic on September 24. Two days later Inez intensified to hurricane status. The storm reached major hurricane status 50 miles east of Guadeloupe on September 27 and then category four status the following day over the northeastern Caribbean. Hurricane Inez crossed over Haiti and Cuba and then turned northeastward over the Bahamas, before reversing direction to the west, resulting in a category one strike at Key West on October 4. Inez returned to category four over the southern Gulf, sideswiping the northwestern corner of the Yucatan Peninsula at Progreso before landfall 35 miles north of Tampico on October 10 as a category three strike.

October 4, 1967: Hurricane Fern; category one strike at Tampico

Fern developed as a tropical storm 350 miles southeast of Tampico on October 2, 1967, and became a minimal hurricane later the same day 300 miles east of Tampico. Fern made landfall on October 4 as a category one hurricane 20 miles north of Tampico. Fern was such a marginal hurricane at landfall that there are no reports of loss of life or damage.

August 23, 1996: Hurricane Dolly; category one strike at Tampico

First identified as a tropical storm over the western Caribbean 150 miles southwest of Grand Cayman on August 19, 1996, Dolly made its first landfall as a minimal hurricane on the Caribbean coast of the Yucatan one day later. After crossing the Yucatan as a tropical depression, Dolly reintensified to category one status on August 23, just before making landfall fifty miles south of Tampico. Like so many hurricanes that come ashore in the vicinity of Tampico, the combination of moist tropical air off the Gulf and the enormous mountain barrier to the west, the Sierra Madre Oriental, produced torrential rains and flooding that in turn caused sudden river flooding along the Panulo River at Tampico.

October 5, 2000: Hurricane Keith; category one strike at Tampico

Keith evolved to a tropical storm over the Bay of Honduras, 250 miles east of Belize, on September 29, 2000. Drifting very slowly toward the northwest, Keith became a hurricane 125 miles northeast of Belize the following day. On October 1, Keith exploded to a category four hurricane, still about 100 miles northeast of Belize. But the next day Keith weakened rapidly to make its first landfall as a category one hurricane in northern Belize. After crossing the Yucatan as a tropical depression, Keith attained minimal hurricane status on October 5, about 150 miles southeast of Tampico, making final landfall as a category one storm the same day 20 miles north of Tampico. Again, there were very heavy rains and flooding in Tampico.

South of Tampico there have been five hurricane strikes at Tuxpan and only one at Nautla, near the northern margins of the Bay of Campeche.

Veracruz, Veracruz State, Mexico

Veracruz, located on the Bay of Campeche about 250 miles southeast of Tampico, is the historic seaport of colonial Mexico, at one time shipping gold and silver on Spanish galleons back to Spain. Today, it is the third largest Mexican city on the Gulf Coast after Tampico and Coatzacoalcos. The population is more than 500,000, with up to 1 million people in the metro region. Sheltered on the Bay of Campeche and with the broad Yucatan Peninsula to the east, Veracruz has suffered only two hurricane

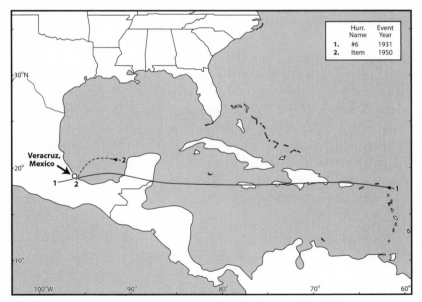

Fig. 6.12. Hurricanes having impact at Veracruz, Mexico, 1901–2007.

strikes since 1901 (see fig. 6.12). In 2007 Hurricane Dean reached category five status over the Caribbean, but weakened thereafter and was never stronger than category one over the Gulf, making landfall as a category one storm near Tuxpan, 150 miles northwest of Veracruz on August 21.

September 16, 1931: storm 6; category one strike at Veracruz

This storm is another Cape Verde Hurricane that originated over the eastern Atlantic on September 8, 1931, and developed to hurricane status east of the Virgin Islands on September 10. After interacting with the mountains of the Dominican Republic, the storm again reached hurricane status off the Yucatan on September 15. The storm made landfall as a very minimal hurricane twenty miles southeast of Veracruz on September 16.

October 10, 1950: Hurricane Item; category one strike at Veracruz

Hurricane Item was a short-lived storm, with its total history over the Bay of Campeche. Item first developed as a tropical storm just off Campeche, 425 miles east of Veracruz. The following day, Item attained minimal hurricane status 250 miles northeast of Veracruz, making landfall twenty miles southeast of the city at the very same location as the hurricane of 1931. Since 1950 there have been no hurricane strikes at Veracruz.

Around the southern shores of the Bay of Campeche there have been extremely few hurricane strikes, even though the bay is considered to be an area of development for tropical storms and hurricanes that then track northward toward Tampico and the Texas coast. There have been no hurricane strikes at Coatzacoalcos and only five tropical storm strikes there. At Ciudad del Carmen there has been one hurricane strike but eleven tropical storms. At Campeche, on the eastern side of the bay and the western coast of the Yucatan Peninsula, there have been four hurricane strikes by hurricanes that have crossed the Yucatan from the Caribbean Sea.

Progreso, Yucatan State, Mexico

Progreso is located near the northwestern corner of the Yucatan Peninsula. It is the seaport for Mérida, the capital city of the state of Yucatan, with a population of about 35,000. Progreso has become the container shipping center for the entire Yucatan Peninsula as well as a primary port for cruise ships with tourists visiting Mérida and Mayan civilization sites. Progreso is sheltered from the onslaught of major hurricanes crossing the Caribbean Sea from east to west by the landmass of the Yucatan Peninsula, with the open waters of the Caribbean at least 175 miles to the east and southeast. Nevertheless, beginning with 1901 there have been twelve hurricane strikes at Progreso, about one every ten years on average. Three of these twelve strikes have been major hurricane strikes. The tracks of the twelve strikes are shown in figure 6.13.

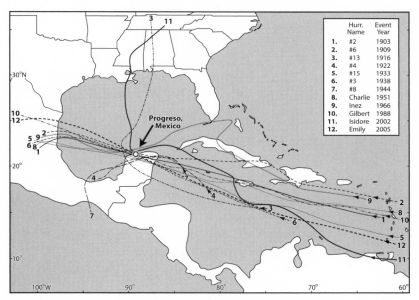

	Hurr. Name	Event Year
1.	#2	1903
2.	#6	1909
3.	#13	1916
4.	#4	1922
5.	#15	1933
6.	#3	1938
7.	#8	1944
8.	Charlie	1951
9.	Inez	1966
10.	Gilbert	1988
11.	Isidore	2002
12.	Emily	2005

Fig. 6.13. Hurricanes having impact at Progreso, Mexico, 1901–2007.

August 13, 1903: storm 2; category one strike at Progreso

This Cape Verde Hurricane was originally identified as a tropical storm midway between Africa and the West Indies on August 6, 1903. It passed close to Martinique as a major hurricane on August 9. Crossing the Caribbean, the hurricane devastated the southern coast of Jamaica and the Cayman Islands before finally making landfall as a category two strike on the Caribbean coast of the Yucatan near Cozumel. The hurricane was a category one strike at Progreso, despite tracking over 175 miles of the Yucatan Peninsula before reaching Progreso. Much of the northern coast of the Yucatan received heavy damage, and some vessels were lost at sea.

August 26, 1909: storm 6; category two strike at Progreso

The sixth storm of the season became a hurricane 300 miles east of Guadeloupe on August 20, 1909. The hurricane interacted with the mountains of Puerto Rico, Hispaniola, and Cuba before reaching major hurricane status again over the western Caribbean, about 200 miles east of Cozumel, on August 25. Later the same day the hurricane made landfall 25 miles south of Cancun as a category two hurricane. After a track of about 150 miles over the Yucatan Peninsula, the hurricane still threatened a category two strike at Progreso as the eye passed the city about 25 miles to the north on August 26. During the hurricane's track over the West Indies and Mexico, heavy rains, floods, and landslides resulted in at least 1,500 deaths.

It was during this hurricane that the first ship-to-shore wireless transmission of weather data was incorporated into weather forecasts. The transmission was from the vessel *Cartago*, unfortunately caught in the hurricane in the Yucatan Channel.

October 16, 1916: storm 13; category two strike at Progreso

The thirteenth storm of this very active season became a hurricane 100 miles south of Jamaica on October 12, 1916. On October 15 the hurricane made landfall as a category two storm on the southern coast of the Mexican territory of Quintana Roo, about 175 miles south of Cozumel. After crossing 200 miles of the Yucatan Peninsula, the hurricane made a category two strike at Progreso, passing directly over the city.

October 19, 1922: storm 4; category one strike at Progreso

First identified as a tropical storm over the western Caribbean 200 miles southeast of Grand Cayman on October 13, 1922, this storm achieved minimal hurricane status on October 17, 175 miles east-southeast of Cozumel. After making landfall at Cozumel on October 18, the hurricane maintained minimal hurricane status during a 200-mile track over the Yucatan Peninsula and then made a category one strike at Progreso, where it passed about 35 miles to the south of the city.

September 23, 1933: Storm 15; category one strike at Progreso

This hurricane was the fifteenth storm of this second-most-active Atlantic Basin season beginning with 1901. The hurricane's development and track are described in the Tampico section. On September 22 the category two hurricane made landfall 25 miles south of Cozumel. After the 200-mile track over the Yucatan Peninsula, the hurricane passed 25 miles south of Progreso, with a category one strike on September 23. This hurricane is especially remembered at Cozumel, where a 300-foot ocean pier and some coastal buildings were destroyed, in addition to several ships.

August 26, 1938: storm 3; category one strike at Progreso

This storm became a hurricane over the western Caribbean 150 miles south of Jamaica on August 23, 1938, with a category two strike just south of Cozumel on August 25. After the 200-mile trajectory over the Yucatan Peninsula the following day, the hurricane passed directly over Progreso as a category one strike. Torrential rains over the peninsula generated massive flooding with nine people drowning in the rushing waters.

September 21, 1944: storm 8; category one strike at Progreso

This storm was first identified as a hurricane over the western Caribbean only 200 miles east of Cozumel on September 19. It made landfall near Cancun the following day as a minimal hurricane. On September 21 the hurricane passed less than 50 miles south of Progreso, as a category one strike.

August 20, 1951: Hurricane Charlie; category two strike at Progreso

The development and track of Charlie are outlined in the Tampico section. Charlie made landfall as a category four strike 25 miles south of Cozumel on August 19, 1951. During Charlie's 200-mile track over the Yucatan Peninsula to Progreso the following day, the hurricane weakened to a category two strike over Progreso.

October 7, 1966: Hurricane Inez; category three strike at Progreso

The development and track of Inez are outlined in the Tampico section. For Progreso in the twentieth century, the trajectory and strike of Hurricane Inez are unique. Inez approached Progreso from the northeast, and the storm center did not cross over the Yucatan Peninsula as with most of the other storms striking this location. The track of category four Inez passed no more than 25 miles to the north of Progreso, with the city positioned in the less dangerous left side of the hurricane.

September 15, 1988: Hurricane Gilbert; category three strike at Progreso

The development and track of Hurricane Gilbert are described in chapter 5. On September 14, 1988, Gilbert made landfall at Cozumel as a category five strike, causing extensive damage to the region. After the 200-mile track across the peninsula, Gilbert was still a major hurricane when it struck Progreso the following day.

October 11, 1995: Hurricane Roxanne; tropical storm strike at Progreso

Hurricane Roxanne briefly reached category three intensity just before landfall on the Caribbean coast south of Cozumel. Roxanne maintained category one status as it slowly crossed the Yucatan Peninsula, with devastating flooding rainfalls in the vicinity of the city of Campeche, but minimal damage at Progreso.

Even though Roxanne was not a hurricane strike at Progreso, it was a notable storm. After passing over Campeche, Roxanne drifted over the Bay of Campeche for six days mostly as a category one hurricane, occasionally as a tropical storm. Roxanne then survived two more days as a tropical depression before losing depression status only 25 miles north of Veracruz on October 20, for a grand total of eight days over the Bay of Campeche!

September 23, 2002: Hurricane Isidore; category three strike at Progreso

Isidore developed as a tropical storm 60 miles south of Jamaica on September 18, 2002, reaching hurricane status the following day 75 miles southeast of the Isle of Youth. Two days later, on September 21, Isidore intensified to a major hurricane only 75 miles northeast of Cancun. During the next two days, Isidore skirted the northern coast of the Yucatan to come ashore on September 23 only 35 miles east of Progreso as a category three strike at that city.

Isidore had been expected to pass to the north of the Yucatan Peninsula, and residents of Progreso and the capital city, Mérida, received only a one-day warning. Nevertheless, there were only a few indirect deaths in the region. However, over the northern half of the peninsula, about 13,000 homes were destroyed, 300,000 people became homeless, and because of torrential rains for thirty-six hours, agricultural losses were enormous. Progreso was badly damaged, and the city was almost a ghost town for months following the hurricane.

July 18, 2005: Hurricane Emily; category one strike at Progreso

Emily is discussed in chapter 5 as one of the memorable hurricanes of the Gulf of Mexico and as one of the four category five hurricanes of the Atlantic Basin 2005 season. Emily was the only one of the four not to be a category five storm over the Gulf. Emily first developed as a hurricane over the Atlantic, unusually far to the south near Tobago. The following day Emily intensified to major hurricane status over the extreme southeastern Caribbean Sea, reaching category five status on July 16 about 100 miles south of Jamaica. On July 18 Emily made landfall in the vicinity of Cozumel as a category four strike, followed by considerable weakening over the Yucatan for a category one strike at Progreso.

Thousands of tourists were evacuated from the Caribbean resorts of Cancun and Cozumel. The sale of alcohol was forbidden in Cancun for thirty-six hours before the predicted landfall of Emily as a precaution for keeping drunken tourists out of harm's way. There was much more storm-related destruction in Cozumel, close to landfall, than in Cancun. The storm surge along the Caribbean coast was reported to be as high as fifteen feet, but there were no deaths associated with the storm on the entire peninsula. Emily was an unusually small hurricane over the Yucatan, and most rainfall recording stations reported less than two inches of hurricane precipitation.

The frequency and intensity of the storm strike record eastward from Progreso along the coast facing the open Gulf of Mexico is similar to that of Progreso. At Rio Lagartos, 100 miles east of Progreso, there have been eleven hurricane strikes, of which four were major, since 1901. This location is still protected from the direct attack of hurricanes approaching the Yucatan from the Caribbean Sea. At Isla Holbox, an additional 75 miles to the east and very close to the northeastern corner of the Yucatan, there have been fourteen hurricane strikes, six of them major hurricane strikes. This site is close to the open waters of the Caribbean to the east and is similarly exposed to the approach of hurricanes off that sea, as are the resort centers of Cancun and Cozumel to the south along the Caribbean coast.

| 7 |

Environmental and Socioeconomic Impacts

Next we examine the impact of hurricanes on the Gulf Coast. Hurricane effects addressed here include population evacuations, heavy rainfall and flooding, storm surges, coastal erosion, and damage to offshore and on-shore oil and gas platforms and related infrastructure. The deadliest and costliest events around the Gulf are also tabulated.

Evacuations

Population along the shores of the Gulf has increased dramatically during the twentieth century as demonstrated in table 7.1. The population in fifty-three coastal counties and parishes from Key West to Brownsville was only 953,000 in 1900. By 1950 the population had increased to 3,751,000, and the last census in 2000 shows the population had exploded to 12,437,000, with many more added since then. At the same time, the tourist industry on the barrier island beaches has increased at a

Table 7.1. Population Statistics for the U.S. Coastal Counties along the Gulf of Mexico in 1900, 1950, and 2000

State	No. of Counties*	1900	1950	2000
Florida	23	175,000	841,000	4,916,000
Alabama	2	76,000	272,000	540,000
Mississippi	3	50,000	127,000	364,000
Louisiana	11	474,000	930,000	1,611,000
Texas	14	178,000	1,581,000	5,006,000
Total	53	953,000	3,751,000	12,437,000

*Parishes for Louisiana

Source: United States Bureau of the Census

Fig. 7.1. Gulf Shores beach resorts.

phenomenal rate (see fig. 7.1). Commercial and industrial development of fisheries, oil and gas, shipping, and other sectors of the regional economy have played a big role as well.

When a hurricane threatens any portion of the coastline, an evacuation of its inhabitants becomes an issue. The dilemma faced by the emergency management community is the classic "crying wolf" syndrome. Emergency management officials can evacuate an area only so often without the area taking a hit before inhabitants no longer heed the request. To further complicate this issue, several regions along the coast require long lead times to evacuate. Currently, our hurricane forecasts simply are not accurate enough to preclude making a mistake or being overly cautious in getting everyone out of harm's way. Obviously, evacuating large urban areas such as Houston, New Orleans, and Tampa requires time simply due to the large volume of people. In addition, there are some remote coastal towns that are connected to more populated areas by a single road, e.g., Key West, where the southern terminus of U.S. Route 1—the Overseas Highway—is 127.5 miles away from the Florida mainland (see fig. 7.2). When a storm threatens this area, everyone must use the same highway off the Keys. The barrier islands along the west coast of Florida, particularly Sanibel and Captiva, adjacent to Fort Myers, and

Fig. 7.2. Overseas Highway, from Key West to Key Largo, Florida.

Longboat Key, between Sarasota and St. Petersburg, have more residents and tourists, but the evacuation routes are much shorter. Many other barrier island examples could be listed. A related but different situation exists for inhabitants of south Louisiana, e.g., Boothville and Venice in southern Plaquemines Parish, Grand Isle in southern Jefferson Parish, or Cocodrie in Terrebonne Parish (see fig. 7.3). Each of these settings is unique—and isolated. Inhabitants in Plaquemines Parish are clustered and exposed along narrow natural levees of the Mississippi River delta that extend for miles. In contrast, Grand Isle is a barrier island, and Cocodrie is simply nestled in the marsh. The city of Galveston, on the barrier island of the same name, faces similar challenges, with only one bridge and a low-lying causeway to the mainland and Houston.

Coordination and cooperation between states are also important to the overall success of any evacuation plan. Contraflow is frequently implemented in these vulnerable areas, where all lanes of traffic are oriented outward from coastal towns and cities. Some recent evacuations have provided test cases for the ability to evacuate large metropolitan

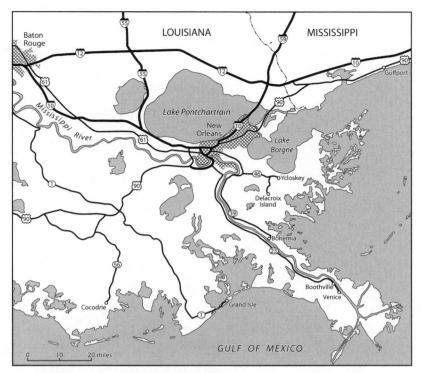

Fig. 7.3. Roads to Boothville-Venice, Grand Isle, and Cocodrie, Louisiana.

areas—Hurricanes Ivan (2004), Katrina (2005), and Rita (2005) (see fig. 7.4). For Ivan, all of the north-central Gulf Coast evacuated concurrently, including the coastal portions of Louisiana, Mississippi, Alabama, and the Panhandle of Florida, leading to gridlock on the freeways. In Louisiana, lessons learned during the false alarm from Ivan led to significant improvements for the evacuation during Katrina one year later. The false alarm probably saved lives during Katrina, the evacuation for which saw a much more successful contraflow operation. There are estimates that over 90 percent of the population of southeastern Louisiana evacuated for this storm, and most of this was accomplished with only about thirty-six hours notice.

The Katrina evacuation and storm aftermath also raised questions about who has the responsibility to call for evacuations. Is it the governor, National Guard, or local officials such as city mayors or parish or county officials? Also, there were questions about how the information should be disseminated, whether over television, radio, or the Internet. Our advice to those in hurricane-prone areas is to acquire information from a variety of sources, being aware that the print media, such as newspapers, cannot keep pace with the rapidly changing details of such operations. Decisions at all levels are vitally important as there are estimates that it costs about $1 million to evacuate one mile of coastline.

Fig. 7.4. Gridlock during the evacuation before Hurricane Rita on I-45 North, heading out of Galveston, Texas.

Heavy Rainfall

Some storms produce incredibly destructive heavy rainfall, while others do not. Overall, the science community does not understand these features well enough to make accurate predictions. Clearly, one factor in rainfall production is the forward speed at which a storm travels. Generally, slow-moving storms produce more rainfall at any given location because of longer exposure to the storm. Unfortunately, that is only one factor among many, most of which are little understood.

Heavy rainfall produced by tropical storms and hurricanes has long been a problem along the Gulf Coast. In addition, there are storms that make landfall along the Gulf Coast, but end up producing heavy rainfall at some distance inland afterwards. For instance, the remnants of Hurricane Camille caused a surprising amount of destructive inland flooding, especially along the foothills of the Blue Ridge Mountains in Virginia, far from the storm's landfall on the Mississippi Gulf Coast.

Among many storms that caused significant inland flooding are Tropical Storm Allison in 2001 and Tropical Storm Isidore in 2002. Tropical Storm Allison is the most costly tropical storm on record, causing approximately $5 billion in damage and twenty-seven fatalities, mostly in southeastern Texas but also in southern Louisiana (see chapter 5). The storm made its initial landfall near Galveston on June 5 and penetrated inland nearly 125 miles. It then moved back south over the Gulf, before moving northeasterly to Morgan City on June 9. It then tracked eastward along the Gulf Coast. Winds with this storm remained below hurricane strength, but rainfall caused much destruction. Hardest hit was Harris County, Texas, which includes the city of Houston. This county experienced Tropical Storm Allison coming and going and had cumulative storm rainfall totals for June 5–9 of over thirty-five inches (see fig. 7.5). The greatest measured rainfall was recorded at the Port of Houston with 36.99 inches, and Thibodaux, Louisiana, recorded 29.86 inches.

Although smaller in scope, rainfall from Tropical Storm Isidore in 2002 also produced impressive rainfall totals. The storm reached its high-

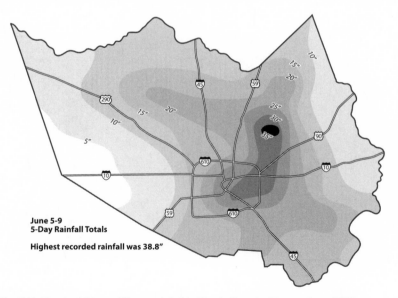

Fig. 7.5. Tropical Storm Allison rainfall in Harris County, Texas, June 2001.

est intensity as a category three storm before making landfall (from the Gulf) on the northern end of the Yucatan Peninsula (see Progreso hurricane history in chapter 6). The storm produced over thirty inches of rain in portions of the Yucatan before turning north toward New Orleans. Heavy rains also fell over the city of New Orleans, with one location within the city recording over twenty-five inches (see fig. 7.6).

Hurricane Camille provides yet another example of heavy rain, though this time the rains were unleashed at some distance from the point of landfall. In this case, the heaviest rains along the coast in Mississippi were close to ten inches, whereas the dying remnants of the storm as it passed over Virginia produced twenty-seven inches in the Appalachians (see fig. 7.7). Since the storm was expected to weaken after landfall, forecasters and the public were largely caught by surprise by the torrential rains. Many rivers in Virginia set records for flood stage during this event. In Virginia alone—not to mention impacts in Mississippi—153 people were killed and over 380 homes destroyed.

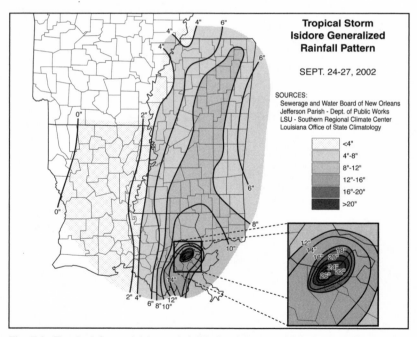

Fig. 7.6. Tropical Storm Isidore rainfall in Louisiana and Mississippi, September 24–27, 2002.

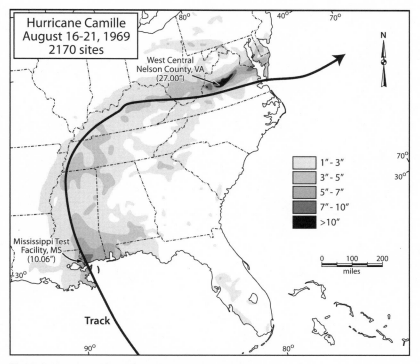

Fig. 7.7. Hurricane Camille rainfall pattern.

Storm Surge

Storm surges result from sea water being pushed against a shoreline by hurricane winds. This abnormal sea level rise is exacerbated by the low atmospheric pressure associated with hurricanes, which allows the sea surface to bulge upward, further contributing to the surge. The bulge contribution to the overall surge, however, is relatively small by comparison. Furthermore, the slope of the affected beach can play a role, whereby a shallow beach slope and gentle slope offshore tend to cause larger surges than a steep coastal slope. Also contributing to local-regional surge is the shape of the coastline. Bays that are open to onshore winds can channel water into confined areas, causing an abnormal surge.

Figure 7.8 shows nine of the cities along the Gulf Coast that are situated on bays or sounds. The bays often provide some protection from wave energy, making these locations desirable for port facilities. How-

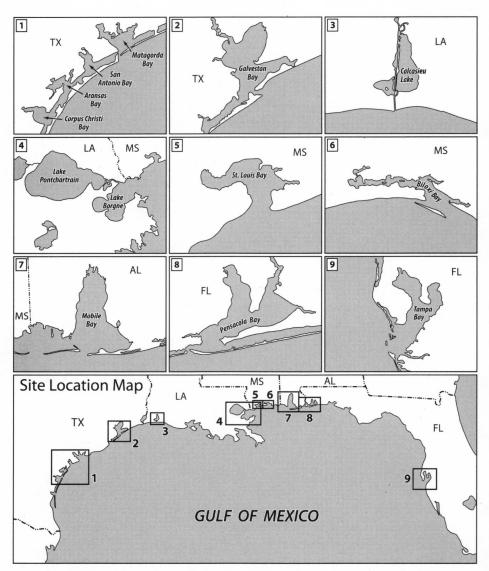

Fig. 7.8. Coastlines at 1) Corpus Christi, Texas; 2) Galveston, Texas; 3) Calcasieu Lake, south of Lake Charles, Louisiana; 4) New Orleans, Louisiana; 5) Bay St. Louis, Mississippi; 6) Biloxi, Mississippi; 7) Mobile Bay, south of Mobile, Alabama; 8) Pensacola, Florida; and 9) Tampa Bay, Florida.

ever, these areas are also vulnerable to storm winds that drive water into the bays. The path of a storm looms large in this equation, as the on-shore winds occur in the right front quadrant of the hurricanes, the eastern side for storms approaching the northern Gulf Coast. Therefore, the portion of coastline most impacted by storm surge is to the right of the storm track. Within this (right-front) quadrant, the forward movement of the storm is additive, with the wind speeds converging inward toward the hurricane's eye. As a result, not only is the surge pushing against the shore in this quadrant of the storm, but the wind speeds are also greater. On the left side of the track, water is generally stripped from the coast, and winds are generally weaker as the forward velocity of the storm serves to reduce wind speeds converging toward the eye.

For example, a worst-case scenario for storm surge in Tampa Bay would be a track southwest–northeast in the Gulf, making landfall just north of the Tampa-St. Petersburg metropolitan area (see fig. 7.8). Hurricane Easy in 1950 (see fig. 6.3) was just such a problematic hurricane for Tampa Bay. In comparison, Galveston Bay and Corpus Christi would likely experience the greatest surge from a storm on a southeast–northwest track, with the storm making landfall just southwest of the city of Galveston (i.e., see Hurricane Jerry track in figure 6.8) or of Corpus Christi. Another interesting setting is Lake Charles, Louisiana, which is actually located about thirty miles inland; however, the Calcasieu Ship Channel and Calcasieu Bay, along with a series of interconnected lakes, provide a conduit for storm surge to reach Lake Charles—both the lake and city. Hurricane Rita in 2005 gives us a great example of this, whereby the storm track was just to the west of Lake Charles and a large surge migrated up the conduit. Storm surge was between fifteen and twenty feet at Cameron, Louisiana, along the coast, but even a six-foot surge reached the I-10 bridge in Lake Charles itself.

Another interesting coastal configuration is the New Orleans metropolitan area. While Hurricane Katrina approached the area from the south, winds over southeastern Louisiana were predominantly from the east, which drove the surge into Lake Pontchartrain and along the levee of the Mississippi River in Plaquemines Parish (see fig. 7.9). The center of Katrina passed about twenty-five miles east of downtown New Orleans and east of Lake Pontchartrain, but easterly winds over the lake gave way to northerly winds, which pushed about nine feet of surge against New Orleans. On the east side of the track—in the right front quadrant of

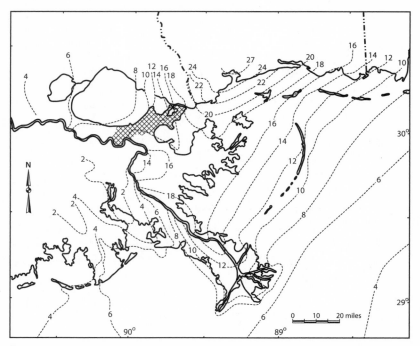

Fig. 7.9. Hurricane Katrina peak storm surge as estimated by the ADCIRC Model.

Katrina—a surge of over twenty-eight feet was recorded in Bay St. Louis, Mississippi. This was the largest storm surge recorded in U.S. history. Interestingly, the area hardest hit by surge in Katrina is not far removed from the area leveled by Hurricane Camille's twenty-three-foot storm surge thirty-six years earlier (see fig. 7.10).

Coastal Erosion

In several locations along the Gulf Coast, land is disappearing. Land gives way to sea for many reasons, including a rising global sea level and, locally and regionally, subsidence of the land surface. In addition, coastal sediments can often be reworked by wave energy, leading to a gnawing away of the coastline. In Louisiana, all three of these factors, plus others, are contributing to disappearing coastal wetlands. The problem is most severe in southeastern Louisiana (see fig. 7.11), where coastal wetlands have been starved of sediment since the levee system was erected

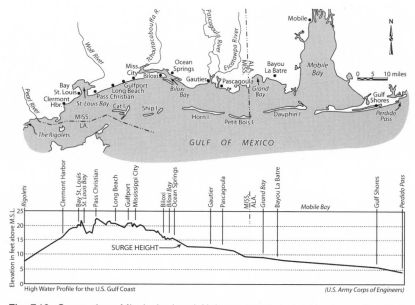

Fig. 7.10. Surge along Mississippi and Alabama coast during Hurricane Camille in 1969.

along the Mississippi River. The river is no longer allowed to flood naturally across the region and leave behind an annual layer of sediment. Now these sediments are often trapped behind dams all along the Missouri-Mississippi-Ohio river basins, and the sediment that does make the trek down the muddy Mississippi is dumped at the edge of the continental shelf, where it travels down the continental slope and rise to the depths (abyssal plain) of the Gulf of Mexico.

This is noteworthy because wetlands serve as a buffer to hurricanes by absorbing much of the wave energy and surge. Estimates from the LSU Hurricane Center suggest that one foot of surge is mitigated for every 3.5 miles of wetlands. As such, if the state of Louisiana could restore thirty-five miles of wetlands between the Gulf of Mexico and the city of New Orleans, for example, ten feet of surge could be absorbed before a storm struck the city. This wetland loss also affects all inland cities along the Gulf and the oil and gas industry, because it allows storms to maintain their intensity at points closer to our major cities. These regions become increasingly vulnerable with each passing year as more and more land gives way to the Gulf.

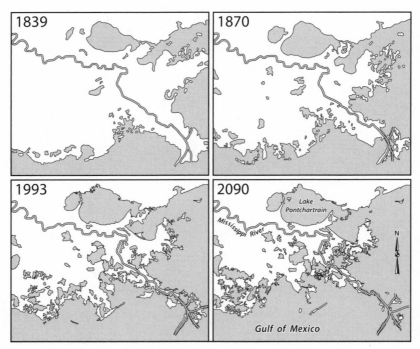

Fig. 7.11. Coastal land loss in Louisiana from 1839 projected to 2090.

Over the last one hundred years, construction in the Louisiana wetlands of thousands of miles of canals for access to oil and gas extraction and processing facilities has also opened up the wetlands for accelerated erosion. This is especially true during tropical storms and hurricanes, when storm waves and surges penetrate much farther inland. The ongoing result is the continued conversion of marshland to open water.

Storm waves and surges are also responsible for the sometimes slow, sometimes rapid displacement of sandy barrier islands fronting the mainland. Virtually the entire coastline of the Gulf is framed with sandy barrier islands that have been intensely developed in recent years with beach homes, condominiums, hotels, and other commercial developments. The waves and surges are responsible for the erosion of beach sands and for at times transporting the sediments over the narrow barrier islands, thereby filling in bays and sounds between the barrier islands and the mainland with the sediment. Various protective structures designed to stabilize current beach positions have been built using local, state, and federal resources, but the long-term effectiveness of these engineering works remains to be seen.

Oil and Gas

There are nearly four thousand oil and gas platforms in the Gulf of Mexico, most of them situated offshore from Louisiana and Texas (see fig. 7.12). During hurricanes, these platforms experience high winds and powerful waves that render them unsafe. As a result, these platforms are mostly evacuated during storms, which is costly to the industry, and are also frequently damaged, as shown in figure 7.13. Furthermore, much of the infrastructure used to convert oil to gasoline is located along the Gulf Coast in proximity to the platforms. These petrochemical plants are also in harm's way when a hurricane strikes, leading to additional damage (see fig. 7.14).

Impacts in this arena reach far and wide, as disruption in the pumping of crude oil and the production of gasoline leads to spikes in pricing. In addition to the higher costs of gasoline and home heating oil, there is a ripple effect in many sectors of the economy—raising transportation costs, storage, and so on.

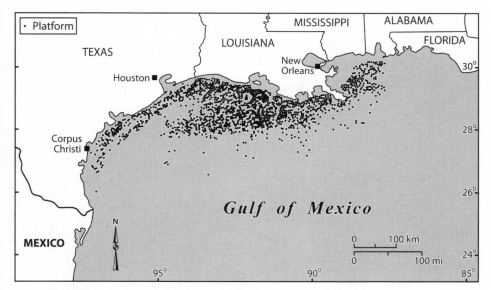

Fig. 7.12. Locations of oil and gas platforms along the U.S. Gulf Coast.

Fig. 7.13. Rig washed aground after Hurricane Katrina, 2005.

Fig. 7.14. Pictures of damage to the oil and gas infrastructure inflicted by Hurricanes Katrina and Rita, 2005.

Most Deadly and Costly Storms

Data for the most deadly and costly storms have been compiled for locations in the United States, but there are few data collected in this regard for Mexico. As a result, we have tabulated the most deadly and costly storms along the U.S. Gulf Coast since 1851. Table 7.2 shows that the most deadly storm was the Galveston Hurricane of 1900. Because of poor forecasting and a number of other factors, over eight thousand people perished, mostly from drowning in the storm surge. The second deadliest storm was Hurricane Katrina in 2005, mainly a result of the precarious setting of New Orleans. Much of the area falls below mean sea level and is protected with a levee system that proved to be inadequate during the storm (see fig. 7.15). Apart from drownings, conditions were so miserable and rescue efforts so slow in the aftermath of the storm that many perished from exposure, heat, and lack of water.

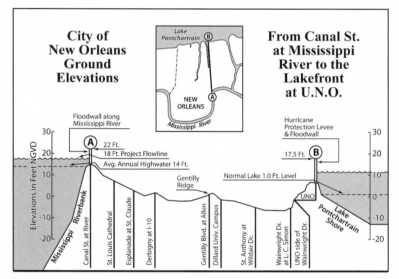

Fig. 7.15. New Orleans in cross-section.

A striking feature of table 7.2 is that eight of the ten deadliest storms occurred in or before 1935, when the population density in this region was significantly lower than in modern times (see table 7.1), This is testament to the fact that in the United States our modern communications and evacuation plans generally work, with occasional breakdowns, such as with Katrina. In contrast, the top ten most costly storms are more recent phenomena, Katrina the most costly by far and Andrew in second place (see table 7.3). Note that Andrew's total includes damage from south Florida on the Atlantic side of the state. All of the most costly storms have occurred in or since 1965, an index of how much infrastructure has increased along the Gulf Coast. This raises questions about changing climates and hurricane regimes, the next topic we will explore.

Table 7.2. Ten Most Deadly Storms along the U.S. Gulf Coast since 1851.

Storm	Date	No. of Deaths
Galveston Hurr.	1900	8,000
Hurr. Katrina	2005	1,500
Cheniere Caminada Hurr. (LA)	1893	1,100–1,400
Hurr. Audrey	1957	416
Great Labor Day Hurr. (Fla. Keys)	1935	408
Isle Derniere Hurr. (LA)	1856	400
FL, MS, and AL	1926	372
LA (Grand Isle)	1909	350
FL Keys/ South TX	1919	287
New Orleans	1915	275
Galveston	1915	275

Source: Blake et al. (2007)

Note: Fatality statistics are difficult to calculate and not all sources agree, e.g., on how to count missing persons, people killed in traffic during an evacuation, etc. The values given in this table are the official estimates as generated by the National Hurricane Center (NHC). Some estimates in the text are from local sources, differing occasionally from this NHC source.

Table 7.3. Ten Most Costly Storms along the U.S Gulf Coast.

Storm	Year	Cost (in billions)
Katrina	2005	84.6
Andrew	1992	48.1
Wilma	2005	21.5
Charlie	2004	16.3
Ivan	2004	15.5
Betsy	1965	11.9
Rita	2005	11.8
Camille	1969	9.8
Frederic	1979	6.9
Allison	2001	6.4

Source: Blake et al. (2007)

| 8 |

THE FUTURE

Are recent Gulf hurricanes—Katrina, Rita, Wilma—a harbinger of the future for the Gulf? Recent scientific inquiry has focused on past hurricane activity in the Gulf and Atlantic to understand past variability and trends and hopefully make future projections. Emerging from these discussions is a lively debate concerning future hurricane climates in the region. At its core, the issue centers on whether global warming has already changed the hurricane climatology of the region. There are two polarized viewpoints on this issue, ranging from global warming extremism on one end to the contrarian (skeptic) point of view on the other, with all shades represented in between. The polarizing viewpoints have created a schism among hurricane researchers, with the dividing line mostly drawn between empiricists and climate modelers. Unfortunately, this debate has grown personal and is fought with a vengeance on both sides.

One side argues that a global warming signal is already detectable in the hurricane records of the Atlantic/Gulf Basin and that modeling experiments suggest increases in the frequency and/or intensity of storms in the future. The other side makes the argument that there is no discernible global warming signal—that our empirical records are inhomogenous over time because our observing platforms have changed over time (see fig. 8.1), and it appears that the historical record of hurricane frequencies and intensities is associated more with shifting sea surface temperatures in the North Atlantic Basin. There are valid points to be made on both sides of the argument.

Regarding the past, the empirical record over the past century or more in the North Atlantic Basin, including the Gulf of Mexico, shows multidecadal scale variability. Figures 3.2 and 3.5 demonstrate this variability through means of coastal strikes along the East and Gulf Coasts, but no discernible trends are visually evident that can be linked to natural or anthropogenic factors. This variability in hurricane activity includes a pe-

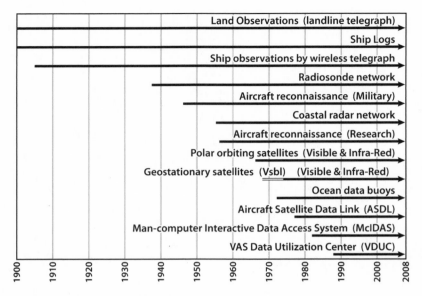

Fig. 8.1. Hurricane-observing platforms over time.

riod of relatively low strike frequencies from the early 1900s to the 1920s, followed by an active period in the late 1920s to the 1960s, and then a period of lower activity in the 1970s through the early 1990s. These multidecadal patterns in hurricane activity appear attributable to shifts in sea surface temperatures (SSTs), where above-normal SSTs in the North Atlantic Basin combined with below-normal SSTs in the southern Atlantic (south of the equator) are most conducive to intense hurricane formation.

These fluctuating SSTs are indexed by the Atlantic Multidecadal Oscillation (AMO) (see fig. 8.2a), which is an index of North Atlantic Ocean SSTs. Often, there is an adjustment made to detrend the AMO data for global warming over this time period (see fig. 8.2b). As shown, the period of high tropical storm and hurricane strikes in the late 1920s–1960s coincides with positive (warm) values of the AMO, while the period of relative calm with regard to strikes in the 1970s–1980s is marked by negative (cool) AMO index values. Specifically in the Gulf, however, these multidecadal patterns are not as distinct as they are for the Atlantic Basin as a whole. The pattern of landfalls indicated in figures 3.2 and 3.5 shows a more steady state of activity over the past century, especially along the

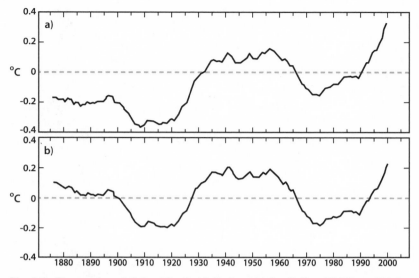

Fig. 8.2. Time series of a) the Atlantic Multi-decadal Oscillation (AMO); and b) the AMO detrended for global warming.

north-central Gulf Coast, with perhaps a more discernible pattern in south Florida as related to the AMO.

Hurricane seasons in 2004 and 2005 were very active along the Gulf Coast, which certainly raises questions about changing hurricane regimes. Over these two seasons, nearly every portion of the north-central Gulf Coast between Galveston, Texas, and Panama City Beach, Florida, was ravaged by a hurricane. Within this more than 500-mile stretch of coastline alone, these two seasons brought Hurricane Ivan and Tropical Storm Mathew in 2004 and Tropical Storm Arlene and Hurricanes Cindy, Dennis, Katrina, and Rita in 2005. There were other major Gulf storms elsewhere over these two years as well, including Hurricanes Charley, Emily, Stan, and Wilma. However, the two seasons following (2006 and 2007) were relatively quiet in this region, with no hurricanes in the Gulf in 2006. Note, however, that no single season, or pair of seasons, really says that much in the context of climate change.

The AMO and SSTs in the North Atlantic do not follow multidecadal patterns in measured surface air temperatures very closely (see figs. 8.3 and 8.4). Global air temperatures show a pattern of increasing values from about 1910 through the early 1940s. From the early 1940s through

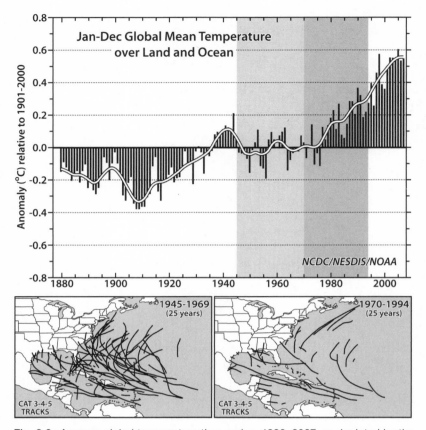

Fig. 8.3. Average global temperature time series, 1880–2007, as depicted by the National Climatic Data Center (*top*) and comparison of severe hurricane tracks over two 25-year periods: 1945–1969, with cool global temperatures and warm SSTs; and 1970–1994, with warm global temperatures and cool SSTs.

the early 1970s, global temperatures remained relatively steady or may have even cooled slightly over this period, before beginning to climb steadily from the late 1970s to the present, with much interannual variability along the way.

It is generally accepted that the hurricane climatology of the North Atlantic Basin has been active since 1995, and many argue that this active period is being driven, at least partly, by warm global temperatures and their impact on SSTs. The opposing viewpoint postulates that the changes in hurricane activity are directly related to the AMO, driven

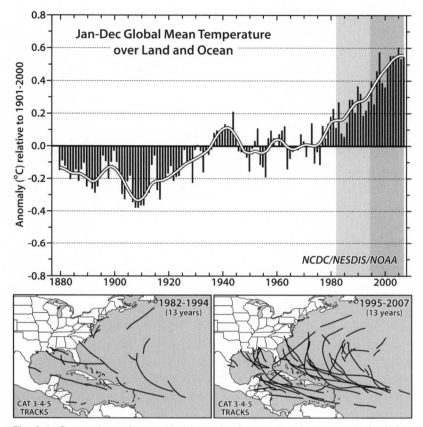

Fig. 8.4. Comparison of severe hurricane tracks over two 13-year periods: 1982–1994, with cool SSTs; and 1994–2007, with warm SSTs.

instead by shifts in the thermohaline (temperature and saline content) circulation of the ocean basins. During periods of strong thermohaline circulation patterns in the North Atlantic, regional SSTs in hurricane breeding grounds tend to be warmer than during periods of weak circulation. As such, Bill Gray and the Tropical Meteorology Project at Colorado State University compare two adjacent periods of twenty-five years; 1945–1969 and 1970–1994 (see fig. 8.3). The former time period is characterized by relative stable global air temperatures that are cooler than current values, but this time period is also represented by strong thermohaline circulation, warm SSTs in the North Atlantic, and a positive AMO. This period is then compared to the following twenty-five years—

1970 to 1994—when global temperatures went into a warming pattern. At the same time, thermohaline circulation was weak and Atlantic SSTs were relatively cool; hence the AMO was negative. A comparison of severe hurricane (category 3–5) tracks between these periods tells a compelling story, whereby the number of tracks in the Atlantic, Caribbean, and Gulf storms is substantially greater during the period with relatively cool global temperatures, despite warm SSTs in the North Atlantic.

Gray and his research group also compare the most recent thirteen-year period (1995–2007)—since the AMO became positive—with the thirteen years prior (1982–1994) (see fig. 8.4). Results are again compelling, showing a dramatic upswing in severe hurricanes during the latter period, corresponding with fluctuating SSTs. Note, however, that the former period includes the intense El Niño of 1982–1983 and the extended period of El Niño conditions from 1990 until early 1995.

Since 1995, every year, with the exception of the El Niño years of 1997 and 2006, had higher-than-average numbers of named storms in the North Atlantic, including 19 named storms in 1995, 16 in 2003, and 15 in 2001, 2004, and 2007, not to mention the record-breaking season of 2005 with 28 named storms. Roughly 10–11 named storms would be considered average for a season. The 1997 season had only seven named storms, coinciding with one of the most intense El Niño events on record. El Niño conditions in the Pacific produce storm-inhibiting westerly wind shear aloft, which is a hostile environment for the formation of Atlantic Basin hurricanes. The 2006 season, which was also an El Niño year, was relatively quiet, with nine named storms. Despite these two quiet seasons, there are indications that the hurricane climatology of the North Atlantic Basin may remain hyperactive for decades to come, based on past AMO history. Using the previous active hurricane period as an analogue from the mid-1920s through the mid-1960s, we could be facing one, two, or perhaps even three decades with mostly above-normal activity during hurricane season, thereby returning hurricane frequencies to what was experienced during the four decades prior to 1965.

Research by Kerry Emanuel suggests that recent hurricanes are associated not only with SSTs, but also with well-documented climate signals, including recent global warming. Since the mid-1970s he has found that the destruction potential of hurricanes has increased markedly due to increased storm lifetimes as well as increased intensities of storms. He postulates that if global temperatures continue to increase, this trend should continue, and will be further exacerbated by increasing human developments in the coastal zone. Note that research in this field does not suggest that global warming will increase the number of storms. In fact, they may decrease in frequency. Emanuel found that storms tend to have longer life spans, and reach higher intensities, thereby producing more major hurricanes.

Regardless of whether global warming is occurring and is affecting hurricane climatology, there are many issues still facing us regarding these powerful storms, as suggested in the last chapter. For example, can we restore wetlands along the Gulf Coast to mitigate hurricane energy? This question looms particularly large in southeastern Louisiana, and Hurricane Katrina inspired renewed consideration of this matter. How will the oil and gas industry in the Gulf cope with future storms? The Gulf is one of that industry's most productive areas in the United States, and storm-related disruptions to offshore drilling, as well as damage to infrastructure, will drive up consumer costs and markedly affect the United States economy. How will we handle continued population growth along the coast, from the perspectives of zoning to minimize damage and evacuations during storm events? How we deal with these and related questions will largely determine how successful we are at living in harmony with our dynamic, though incredibly productive, coastline.

Epilogue

ANOTHER DISASTROUS SEASON: 2008

Following the destructive 2004 and 2005 hurricane seasons on the Gulf Coast, those living along or near the Gulf shores were justifiably concerned about the coming hurricane seasons, especially since most seasonal hurricane forecasts called for greater-than-normal activity for the entire Atlantic Basin for years, and perhaps decades, to come. Although the 2006 season was near average in terms of tropical storms, hurricanes, and major hurricanes over the Atlantic Basin, there were only two tropical storm landfalls along the coasts of the Gulf: Tropical Storm Alberto, with a landfall in a rural area east of Apalachicola, Florida, and Hurricane Ernesto, which made landfall as a tropical storm in the Florida Keys northeast of Islamorada, before passing over the Gulf waters of Florida Bay, with a second landfall near Flamingo on the Florida mainland. There was a general sense of relief and gratitude for only two relatively weak storms that caused little damage, and Gulf Coast residents were hopeful that the pattern would continue despite the ongoing seasonal forecasts.

For the entire Atlantic Basin, the 2007 season returned to above normal activity, with fifteen named storms, including six hurricanes, two of which, Dean and Felix, reached category five status. Two tropical storms and three hurricanes were over Gulf waters. Tropical Storm Barry struck along the west coast of Florida, and Erin struck along the south Texas coast near Corpus Christi. The season had three category one hurricane strikes: Hurricane Dean two times on both sides of the Bay of Campeche, including the city of Campeche on the east and Tuxpan on the west; Hurricane Humberto in the vicinity of Galveston; and Hurricane Lorenzo again in the vicinity of Tuxpan. Relative to 2004 and 2005, however, these storms were locally pesky and dangerous, but not in the same class as the memorable events of two and three years earlier.

Again, for the Atlantic Basin, the 2008 season was above average, with fifteen named storms, seven of which reached hurricane status, and four of those becoming major hurricanes—Bertha, Gustav, Ike, and Omar. Around the Gulf of Mexico, three were tropical storm strikes, Edouard near the Texas-Louisiana border on August 5; Fay, with multiple strikes around the eastern and northeastern Gulf coasts at Key West, Marco Island, Cedar Key, Apalachicola, Panama City, and Destin between August 19 and 23; and Marco at Nautla, north of Veracruz, Mexico, on October 7.

Hurricane Dolly came on shore as a category two strike in the resort areas of South Padre Island on July 23 as the first hurricane strike of the season in the United States. Dolly was initially designated as a tropical storm over the western Caribbean on July 20, making its first landfall on the Yucatan Peninsula the following day. In Guatemala, seventeen people lost their lives, twelve from a single family. Dolly intensified over the Bay of Campeche, causing about $1.2 billion in damage along the upper Mexican coast in the state of Tamaulipas, around Brownsville and South Padre Island, and especially over the Lower Rio Grande Valley, where the losses were primarily agricultural as a result of heavy rainfall and flooding. Along the beaches the effects of hurricanes can be extremely far reaching. For example, one person drowned and nine were rescued from strong rip currents believed to be associated with Hurricane Dolly at Panama City Beach in Florida, about eight hundred miles northeast of landfall in Texas.

Unfortunately, however, there were two extremely destructive hurricane strikes along the northwestern Gulf Coast later in the 2008 season. The first was Hurricane Gustav as a strong category two strike in southern Louisiana on September 1, and then Hurricane Ike, another category two strike in southeastern Texas on September 13. Both of these hurricanes were far more catastrophic in terms of impact than their category two classification at landfall would imply. We will focus now on these two devastating hurricanes, thus ending this book where we began— with a severe hurricane at Galveston, occurring 108 years after the 1900 storm.

Tropical Storm Gustav was first identified on August 25 over the central Caribbean Sea, 175 miles south of the Dominican Republic. Gustav moved northwestward and intensified to hurricane one status the following day as it made landfall in Haiti. After interacting with the high mountains of southwestern Haiti, Gustav turned abruptly to the southwest over the open Caribbean, now as a weaker tropical storm following a zigzag track across Jamaica. Gustav began to intensify again late on August 29 after departing from Jamaica. The storm swept northwestward, passing to the northeast of the Cayman Islands as a category one hurricane before intensifying rapidly as it crossed the Isle of Youth near western Cuba. Gustav made landfall as a category four storm in the province of Pinar del Rio, where about 100,000 dwellings were damaged or destroyed. Gustav was declared the most damaging storm in Cuba since Flora in 1963, although there were no reports of storm-related deaths there. However, elsewhere in the Carribbean, the death toll included eight in the Dominican Republic, seventy-six in Haiti, and eleven in Jamaica, in addition to much destruction and misery wrought by the storm prior to its landfall in western Cuba.

After leaving Cuba, Gustav continued on a relatively straight track across the Gulf toward the northwest as a category three hurricane. Gustav made landfall as a borderline category two or three hurricane (officially a two) on September 1 in southeastern Louisiana at Cocodrie, about sixty miles southwest of New Orleans.

As Gustav swept over the Gulf on August 31, threatening southern Louisiana and adjacent coastal Mississippi, the National Hurricane Center forecast a category two-to-three hurricane strike. Mindful of Katrina events only three years earlier, New Orleans mayor Ray Nagin ordered a mandatory evacuation of the city on August 30 at 8:45 p.m. CDT, two days prior to the anticipated landfall. He called Gustav the "storm of the century and the mother of all storms" to motivate residents to act, as reported by the *New York Times* on the following day. CNN also quoted Nagin as saying, "You need to be scared, you need to be concerned, and you need to get your butts moving out of New Orleans right now." Altogether about 1.9 million people evacuated southern Louisiana, including almost everyone from New Orleans, making this the largest and most orderly evacuation in Louisiana history. New Orleans residents were given the official green light to return five days later, on September 4, even though electrical power was not yet fully restored in much of the city. Given the normal uncertainties of hurricane forecasting, there were evacuations, both voluntary and mandatory, along a five-hundred-mile coastline from Mobile on the east to near Galveston on the west.

Fortunately for New Orleans, the city managed to "dodge the bullet" once again, a maneuver that has failed only once—Hurricane Katrina—

since 1965. Over New Orleans, Gustav was only at tropical storm strength, and the partially rebuilt and reinforced levees were not breeched, even though they were briefly overtopped by waves, particularly along the Industrial Canal across from where the levees were breached in the Lower Ninth Ward during Katrina. The moveable barriers where three drainage canals empty into Lake Pontchartrain also were effective in preventing the higher levels of the lake from entering the canals and flooding low-lying areas of the city. U.S. Corps of Engineers officials, responsible for these new flood protection structures, were especially relieved, but at the same time cautioned that much more costly work is needed if New Orleans is to be protected from a hundred-year hurricane strike. The city sustained expectable damage of fallen branches and trees, removal of shingles and roofs, and street flooding.

Closer to the Gulf, hurricane winds and the storm surge were particularly destructive in Plaquemines, Lafourche, and Terrebonne Parishes, including the city of Houma. These parishes are economically dependent on the offshore oil and gas industry, the seasonal harvests of shrimp, oysters, and crabs distributed to local and national markets, and commercial and sport fishing over the open Gulf.

The fate of Port Fourchon, fifty miles south of New Orleans, on a spit of slightly higher ground between Bayou Lafourche and the Gulf of Mexico, is of special concern during hurricane strikes across southeastern Louisiana. From 16 to 20 percent of the nation's oil and gas pass through this industrial seaport on the very edge of the Gulf. About 270 ships a day service 90 percent of the 3,500 offshore oil and gas platforms through Port Fourchon. It is also the site of the Louisiana Offshore Oil Port, LOOP, where supertankers transfer their products directly to pipelines connecting to the mainland. The narrow two-lane Louisiana Highway 1 connecting Port Fourchon across seventeen miles of eroding saltwater marsh, with slightly higher ground at Golden Meadow, is the only vehicle access for workers and supplies. A seven-mile elevated highway from Port Fourchon north to Leeville is scheduled to be completed in 2009, and there are plans for an additional ten miles of elevated highway to Golden Meadow sometime in the future.

The continued functioning of the oil and gas facilities at Port Fourchon during and after hurricanes is most important for the U.S. economy. Hurricane Gustav made landfall only about twenty miles to the west of Port Fourchon, and the damage to essential facilities was much less than feared. Nevertheless, full electrical power was not restored for more than one week, and the facilities had to operate at reduced capacity using diesel generators. The ship channels were clogged with uncharted debris, and the highway back to Golden Meadow was covered by a thick coating of oily sediment sludge washed up by the six- to seven-foot storm surge. Most of the seafood-processing establishments along Bayou Lafourche and the commercial and residential neighborhoods farther inland were heavily damaged by wind, and some buildings floated off their foundations due to the storm surge and local flooding. Devastation associated with Hurricane Gustav was not as great as feared, but one week later, residents, workers, and officials were already concerned about the threatened arrival of Hurricane Ike over the southeastern Gulf of Mexico, with the likelihood of another hurricane landfall in Louisiana or eastern Texas in several days.

Probably most unexpected was the impact of hurricane-force winds, two peak gusts of ninety-one miles per hour in Baton Rouge, more than sixty miles inland from the Gulf of Mexico. Baton Rouge is recognized for its massive tree canopy. Red and water oaks, sycamores, and pines were particularly vulnerable to the hurricane-force gusts, and falling limbs and massive oaks crashed through homes, blocked streets, and brought down electrical lines across the city (see fig. E-1). By the time Gustav departed, almost the entire city was without power, and two weeks later there still were neighborhoods where power had not been restored. Property losses in Baton Rouge were much greater than for Hurricane Betsy in 1965 and Hurricane Andrew in 1992, the two most destructive and memorable hurricanes during the latter half of the twentieth century in the city. The task of removing and disposing of broken and downed trees—sometimes with trunks six or more feet in diameter—seemed endless, with head-high rows of debris giving a canyonlike appearance to many streets and roads weeks after the hurricane event. The national media had been positioned to report again about the impending disaster in New Orleans, and when Gustav bypassed New Orleans, very little national attention was given to the almost-paralyzing situation in Baton Rouge. Tropical Storm Gustav was responsible for 76 deaths in Haiti and 43 in the United States, and about $8.3 billion in damage.

The tropical disturbance that later became Hurricane Ike was first identified as a tropical storm over the mid-Atlantic, west of the Cape Verde Islands, on September 1. On September 4, over the open Atlantic

Fig. E-1. Pin oak damage to the home of Woody Keim in Prairieville, Louisiana, during Hurricane Gustav.

northeast of Puerto Rico, Ike intensified from category one to four in only six hours, passing to the north of Haiti. There the storm generated massive floods around Gonaïves and caused seventy-four deaths. In only three weeks, Haiti had been devastated by four tropical storms and hurricanes—Fay, Gustav, Hanna, and Ike. After crossing the Turks and Caicos Islands three days later on September 7 as a category three to four hurricane, Ike made its first landfall in eastern Cuba at Holguin Province, as a category three storm. Ike then crossed Cuba and followed the southern coast as a category one hurricane, making a second landfall in Pinar del Rio province, where Gustav had ravaged the region only ten days earlier. The sugarcane crop was destroyed, and total damages from Gustav and Ike combined were estimated at more than $5 billion in U.S. dollars The storm season was declared to be the worst ever in Cuba.

Hurricane Ike then exited Cuba on Tuesday, September 9, and began to move over the eastern Gulf of Mexico in the general direction of Texas or Louisiana, initially as a category one hurricane. Two days later Ike was over the central Gulf as a very large category two hurricane, with the potential to intensify to a category three and possibly even to four over the very warm Gulf waters, before coming ashore with a massive destructive storm surge. On Friday, September 12, Ike stubbornly refused to intensify to category three, but the forecasts were for landfall the next day near the western end of Galveston Island. The National Weather Service warned that people who remained in one- or two-story homes on Galveston Island and the Bolivar Peninsula immediately to the east would not survive the storm surge, which was predicted to reach 18 to 25 feet. About 2.2 million residents evacuated from threatened neighborhoods of southeastern Texas and an additional 130,000 from southwestern Louisiana. Nevertheless, about 140,000 residents remained in areas where mandatory evacuations were ordered, including people in the very threatened areas of Galveston Island and Houston and an unknown number on the extremely risky Bolivar Peninsula.

Ike made landfall the following morning as a category two hurricane near the eastern end of Galveston Island. The last-minute jog to the right somewhat diminished the destructive forces of Ike along the unprotected but developed West End of Galveston Island and the petrochemical complexes a little farther to the west at Freeport. However, the jog intensified the potential for disaster along the Houston Ship Channel, the greatest concentration of petrochemical complexes along the Gulf Coast. Threats increased for the beach resort communities eastward along the Bolivar Peninsula, the petrochemical facilities at Beaumont and Port Arthur, and even beyond across coastal Louisiana to New Orleans.

Across southeastern Texas and southwestern Louisiana, hurricane and gale-force winds took down trees and wires. Three to four million inhabitants lost electrical power, with many waiting for restoration two weeks or more after landfall. In much of the impacted region away from the coast, most of the economic uncertainties and losses and emotional stress were caused by fallen trees and flash flooding blocking streets and highways and the extended loss of electrical power. In downtown Houston, for example, shards of glass rained down from skyscraper windows, and some buildings were affected by flash flooding. Surprisingly, however, there was minimal damage to the petrochemical complexes along the Houston Ship Canal.

At Galveston and on the Bolivar Peninsula, on the other hand, Ike caused damage and destruction that will take months or even years to repair (see fig. E-2). The storm surge at Galveston, which was predicted to reach 15 to 25 feet, only averaged about 12 feet. Still, the entire island was flooded by sea water, with historic buildings in the downtown, known as the Strand, flooded to a depth of six feet. Unlike in 1900 when city blocks near the Gulf were swept clean, most buildings were damaged but still standing, with furniture and belongings swept into piles of debris imbedded in a residue of sand and silt. Seven buildings near the sea wall collapsed, and ten more were destroyed by fire when firemen were unable to reach them during the storm. Newer high-rise condominiums adjacent to the beach at the eastern end of the island, very close to the site of landfall, survived with minimal damage.

The thirty-mile-long Bolivar Peninsula, similar to a narrow barrier island but connected to the mainland at High Island, was almost swept clean, much like the Mississippi Gulf Coast during Katrina in 2005. Some residents who planned to evacuate apparently did not realize that destructive winds and storm surge would arrive well in advance of the predicted hurricane landfall when the central eye reaches the coast. They waited too long, and could not escape after the storm surge flooded the only coastal road to High Island and the highway inland to Winnie. Some survived in their homes or other shelters, but others did not, and weeks later several hundred residents were still listed as missing.

Immediately after the departure of Hurricane Ike, there was no power

Fig. E-2. Hurricane Ike damage on the Bolivar Peninsula at Gilchrist: a) lone house rising above the debris; b) most houses swept off their foundations.

or water and no sewer services on Galveston Island and the Bolivar Peninsula. On Wednesday, September 17, four days after landfall, Galveston residents were allowed to return briefly to survey their homes and retrieve valuables and some essentials, but the operation had to be canceled later the same day because of debris-filled streets and gridlock situations on the only access bridge to Galveston. Residents were finally allowed to return on Wednesday, September 24, thirteen days after mandatory evacuation and eleven days after landfall, still with limited power, water, and sewer services and a twelve-hour overnight curfew. There was still no provision for access of Bolivar Peninsula residents to their mostly destroyed properties.

Out over the Gulf, Hurricane Ike was geographically so large that significant storm surges were experienced eastward across Louisiana to the Mississippi River delta. South of Lake Charles, Louisiana, the surge swept inland nearly as far as the surge from Hurricane Rita three years earlier in 2005, and to the east in Terrebonne Parish south of New Orleans, 13,000 homes were flooded in the vicinity of Houma.

Hurricane Ike was responsible for thirty-seven deaths in Texas and eight in Louisiana, with more than two hundred residents missing. Total economic losses in the United States are estimated to be as high as $27 billion, making Ike the third mostly costly hurricane in United States history after Katrina in 2005 and Andrew in 1992. These numbers are reminders that future losses will continue to be high given the almost "wall-to-wall" residential and commercial developments of the barrier islands and adjacent coastal regions.

Many of the oil and natural gas production platforms in the Gulf were shut down before Gustav and Ike swept across the Gulf. On September 23, the Minerals Management Service (MMS) reported that forty-nine platforms were destroyed in the Gulf, but that they represented only 1 percent of production from the Gulf. At that time, six refineries in Texas were still not operating, but these represented less than 10 percent of United States capacity. Production was back to normal by late November.

Full recovery after Ike in Galveston and on the Bolivar Peninsula will be a long time coming. On October 13, there were still more than eleven thousand roofs with blue tarps between Galveston and the Louisiana border. Water, sewer, and electrical service had been restored to most of Galveston, but access to the Boliver Peninsula was extremely limited. Full access was reopened before the holiday season in December. Thankfully, there were no more hurricane strikes around the Gulf during the remainder of the 2008 season, though we are likely to face hyperactive hurricane seasons for the foreseeable future.

Sources and Suggested Readings

Anthes, R. 1982. *Tropical Cyclones: Their Evolution, Structure, and Effects.* Boston, MA: Meteorological Monographs, Vol. 19, American Meteorological Society.

Barnes, J. 2007. *Florida's Hurricane History,* 2nd ed. Chapel Hill, NC: University of North Carolina Press.

Bechtel, S. 2006. *Roar of the Heavens: Surviving Hurricane Camille.* New York: Citadel Press.

Beven, J. L., II. "Blown Away: The 2005 Atlantic Hurricane Season." *Weatherwise,* July-August, 2006, 32–44.

Blake, E. S., Rappaport, E. N., and Landsea, C. W. 2007. *The Deadliest, Costliest, and Most Intense United States Tropical Cyclones from 1851 to 2006 (and Other Frequently Requested Hurricane Facts).* National Weather Service, National Hurricane Center, Miami, Florida.

Bourne, J. K., Jr. 2007. "Should New Orleans Rebuild?" *National Geographic Magazine,* August 2007, pp. 32–67.

Bove, M. C., J. B. Elsner, C. W. Landsea, X. Niu, and J. J. O'Brien. 1998. "Effect of El Niño on U.S. Landfalling Hurricanes Revisited." *Bulletin of the American Meteorological Society* 79, 2477–2482.

Brinkley, D. 2006. *The Great Deluge.* New York: Harper Collins.

Campanella, R. 2002. *Time and Place in New Orleans: Past Geographies in the Present Day.* Gretna, La.: Pelican Publishing Co.

Carrier, J. 2001. *The Ship and the Storm: Hurricane Mitch and the Loss of the Fantome.* Camden, Maine: International Marine/McGraw-Hill.

Cry, G. W., and W. H. Haggard. 1962. "North Atlantic Tropical Cyclone Activity, 1901–1960." *Monthly Weather Review* 90: 341–349.

Curry, J. A., P. J. Webster, and G. J. Holland. 2006. "Mixing Politics and Science in Testing the Hypothesis That Greenhouse Warming Is Causing a Global Increase in Hurricane Intensity." *Bulletin of the American Meteorological Society* 87:1025–1037.

Dean, C. 1999. *Against the Tide: The Battle for America's Beaches.* New York, NY: Columbia University Press.

Diaz, H., and R. Pulwarty. 1997. *Hurricanes: Climate, Social, and Economic Impacts*. New York, NY: Springer-Verlag.

Dixon, T. H., et al. 2006. "Space Geodesy: Subsidence and Flooding in New Orleans." *Nature* 441, doi:10.1038/441587a, pp. 587–588.

Doehring, F., I. W. Duedall, and J. M. Williams. 1994. *Florida Hurricanes and Tropical Storms, 1871–1993: An Historical Survey*. Gainesville, FL: Florida Sea Grant College Program TP-71, University of Florida.

Drye, Willie. 2002. *Storm of the Century: The Labor Day Hurricane of 1935*. Washington, D.C.: National Geographic Society.

Dunn, G. E., and B. I. Miller. 1964. *Atlantic Hurricanes*. Baton Rouge, LA: Louisiana State University Press.

Elsner, J. B., and B. Kara. 1999. *Hurricanes of the North Atlantic: Climate and Society*. New York: Oxford University Press.

Emanual, K. 2005. "Increasing Destructiveness of Tropical Cyclones over the Past 30 Years." *Nature* 436:doi:10.1038/nature03906.

Emanuel, K. 2005. *Divine Wind: The History and Science of Hurricanes*. New York, NY: Oxford University Press.

Gray, W. M. 1999. *On the Causes of Multi-Decadal Climate Change and Prospects for Increased Atlantic Basin Hurricane Activity in Coming Decades*. 10th Symposium on Global Change Studies, Preprints, American Meteorology Society, pp. 183–186.

Greene, C. E., and S.H. Kelly, eds. 2000. *Through a Night of Horrors: Voices from the 1900 Galveston Storm*. College Station, TX: Texas A&M University Press.

Grunwald, M. 2007. "The Threatening Storm." *Time*. Aug. 13, pp 30–39.

Halverson, J. B. 2006. "A Climate Conundrum." *Weatherwise*, March–April, 2006, 19–23.

Hearn, P. 2004. *Hurricane Camille: Monster Storm of the Gulf Coast*. Jackson, MS: University Press of Mississippi.

Henderson-Sellers, A., H. Zhang, G. Berz, K. Emanuel, W. Gray, C. W. Landsea, G. Holland, J. Lighthill, S-L. Shieh, P. Webster, and K. McGuffie. 1998. "Tropical Cyclones and Global Climate Change: A Post-IPCC Assessment." *Bulletin of the American Meteorological Society* 79, 19–38.

Henry, W. K., D. M. Driscoll, and J. P. McCormack, 1975. *Hurricanes on the Texas Coast*. College Station, TX: Texas A&M University Press.

Horne, J. 2006. *Breach of Faith: Hurricane Katrina and the Near Death of a Great American City*. New York, NY: Random House.

Ike: Stories of the Storm. 2008. Galveston, TX: Galveston County Daily News.

Keim, B. D., R. A. Muller. 2008. "Overview of Atlantic Basin Hurricanes," chapter 4 in Patrick J. Walsh, ed., *Oceans and Human Health: Risks and Remedies from the Seas*. Burlington, MA: Academic Press/Elsevier.

Keim, B. D., R. A. Muller, and G. W. Stone. 2004. "Spatial and Temporal Variability of Coastal Storms in the North Atlantic Basin." *Marine Geology* 210, 7–15.

Keim, B. D., R. A. Muller, and G.W. Stone. 2007. "Spatiotemporal Patterns and Return Periods of Tropical Storm and Hurricane Strikes from Texas to Maine." *Journal of Climate* 20(14): 3498–3509.

Keim, B. D., and K. D. Robbins. 2006. "Occurrence Dates of North Atlantic Tropical Storms and Hurricanes: 2005 in Perspective." *Geophysical Research Letters* 33, L21706, doi:10.1029/2006GL027671.

Landsea, C. W. 2007. "Counting Atlantic Tropical Cyclones Back to 1900." *EOS*, Vol. 88, No. 18, p. 197.

Landsea, C. W. 2005. "Hurricanes and Global Warming." *Nature* 438:E11–E13.

Landsea, C. W., R. A. Pielke, Jr., A. M. Mestas-Nunez, and J. A. Knaff. 1999. "Atlantic Basin Hurricanes: Indices of Climate Changes." *Climatic Change* 42, 89–129.

Larson, E. 1999. *Isaac's Storm.* New York, NY: Crown Publishers.

Ludlam, D. M. 1963. *Early American Hurricanes, 1492–1870.* Boston, MA: American Meteorological Society.

McComb, D. G. 1986. *Galveston: A History.* Austin, TX: Univ. of Texas Press.

Maklansky, S. 2006. *Katrina Exposed: A Photographic Reckoning.* New Orleans: New Orleans Museum of Art.

Marasalis, W. 2005. *Hurricane Katrina: The Storm That Changed America.* New York: Time Books.

Mooney, C. 2007. *Storm World: Hurricanes, Politics, and the Battle over Global Warming.* Orlando, FL: Harcourt.

Moyer, S. M., ed. 2005. *Katrina: Stories of Rescue, Recovery, and Rebuilding in the Eye of the Storm.* Champaign, IL: Spotlight Press

National Geographic Magazine, August 2006. Three articles: "Home No More," photo essay by David Burnett, 42–53. "Where Have You Gone, New Orleans?" by Ernest J. Gaines, 54–65. "Super Storms: 'No End in Sight,'" by Thomas Hayden, 66–77.

National Hurricane Center, Catalog of Atlantic Basin tropical storm and hurricane tracks, 1851 to date available at http://weather.unisys.com/hurricane/atlantic/index.html and their archive of real-time advisories and storm summaries at http://www.nhc.noaa.gov/pastall.shtml .

Neumann, C. J., B. R. Jarvinen, C. J. McAdie, and J. D. Elms. 1993. "Tropical Cyclones of the North Atlantic Ocean, 1871–1992." Historical Climatology Series 6-2, National Climatic Data Center, Asheville, NC.

Norcross, B. 2006. *Hurricane Almanac 2006: The Essential Guide to Storms Past, Present, and Future.* New York: St. Martin's.

Pielke, R. A., Jr., and R. A. Pielke, Sr. 1997. *Hurricanes: Their Nature and Impacts on Society.* New York: John Wiley and Sons.

Pielke, R. A., Jr., C. W. Landsea, M. Mayfield, J. Laver, and R. Pasch. 2005. "Hurricanes and Global Warming." *Bulletin of the American Meteorological Society* 86:1571–1575.

Post, C. 2007. *Hurricane Audrey: The Deadly Storm of 1957.* Gretna, LA: Pelican Publishing Co.

Rose, C. 2007. *1 Dead in the Attic: After Katrina.* New York: Simon & Schuster.

Rosenfeld, J. 2005. "The Mourning after Katrina." *Bulletin of the American Meteorological Society* 86:1555–1566.

Scott, P. 2006. *Hemingway's Hurricane: The Great Florida Keys Storm of 1935.* Camden, Maine: International Marine/McGraw Hill.

Shangle, R. D. 2005. *The Wrath of Hurricane Katrina.* Beaverton, OR: American Products Publishing Co.

Simpson, R. H., and M. B. Lawrence. 1971. *Atlantic Hurricane Frequencies Along the U.S. Coastline.* Silver Springs, MD: NOAA Technical Memorandum NWS SR-58.

Simpson, R. H., and H. Riehl. 1981. *The Hurricane and Its Impact.* Baton Rouge, LA: Louisiana State University Press.

United States Corps of Engineers.1970. *Hurricane Camille, 14–22 August 1969.* Mobile, AL.

Van Heerden, I., and M. Bryan. 2006. *The Storm.* New York: Viking.

Vermani, J. I., and R. H. Weisberg. 2006. "The 2005 Hurricane Season: An Echo of the Past or a Harbinger of the Future?" *Geophysical Research Letters* 33, L16705, doi:10.1029/2006GL02869.

Weems, J. E. 1957. *A Weekend in September.* New York: Henry Holt and Co.

Wells, K. 2008. *The Good Pirates of the Forgotten Bayous: Fighting to Save a Way of Life in the Wake of Hurricane Katrina.* New Haven, CT: Yale University Press.

Zebrowski, E., and J. Howard. 2005. *Category 5: The Story of Camille, Lessons Unlearned from America's Most Violent Hurricane.* Ann Arbor: University of Michigan Press.

ILLUSTRATION CREDITS

Figures not listed below were prepared by the Cartographic Information Center of the Department of Geography and Anthropology at Louisiana State University.

1. THE GALVESTON HURRICANE OF 1900

Fig. 1.2. Wagon bridge connecting Galveston with mainland is reproduced courtesy Rosenberg Library, Galveston, Texas.

Fig. 1.3. Storm track and Saffir-Simpson intensity categories of the Galveston Hurricane of 1900 are as interpreted by the National Hurricane Center.

Fig. 1.4. Photo of Dr. Isaac Cline is reproduced courtesy NOAA/Department of Commerce Photo Library.

Fig. 1.5. Photo of relief workers searching for bodies is from the Library of Congress (LC-USZ62-71880).

Fig. 1.6. Generalized map of Galveston after the hurricane was redrafted with permission from a map in the Rosenberg Library collection, Galveston.

Fig. 1.7. Photo of Galveston residential area is reproduced courtesy NOAA/Department of Commerce Photo Library. Photo of debris wall is reproduced courtesy Rosenberg Library, Galveston. Photo of remaining walls of Sacred Heart Church is reproduced courtesy Rosenberg Library, Galveston. Photo of house lifted off its foundation is from the Library of Congress.

Fig. 1.8. Photo of removal of bodies from Galveston landscape is reproduced courtesy NOAA Photo Library.

Fig. 1.9. Photo of Galveston's sea wall and promenade is reproduced courtesy Rosenberg Library, Galveston.

Fig. 1.10. Photo of Galveston sea wall in more modern times was taken by Robert Muller.

Fig. 1.11. Photo of slurry pumped through pipes is reproduced courtesy Rosenberg Library, Galveston.

Fig. 1.12. Photo of Galveston residences and land being raised is reproduced courtesy Rosenberg Library, Galveston.

Fig. 1.13. Photo of St. Patrick's Church raised with jacks is reproduced courtesy Rosenberg Library, Galveston.

2. HURRICANE KATRINA

Fig. 2.2. Analysis, map, and graph by Richard Campanella, *Geographies of New Orleans: Urban Fabrics Before the Storm* (Lafayette: Center for Louisiana Studies, 2006), 55, used with permission.

Fig. 2.3. Storm track of Hurricane Katrina is courtesy National Hurricane Center.

Fig. 2.4. Forecasted 5-day track for Hurricane Katrina is courtesy National Hurricane Center.

Fig 2.5. Chart of location of levee failures in New Orleans is adapted from T. H. Dixon et al. (2006).

Fig 2.6. Photo pair of the Biloxi, Mississippi, coast is used courtesy of the United States Geological Survey.

Fig 2.7. Photo pair of the Waveland, Mississippi, coast is used courtesy of the United States Geological Survey.

Fig. 2.8. Underlying data for generalized flood depths of New Orleans on 31 August 2005 are from NOAA at http://www.katrina.noaa.gov/maps/maps.html.

Fig. 2.9. Photo of roof surface peeled off of Superdome is used courtesy of the National Oceanographic and Atmospheric Administration from www.katrina.noaa.gov/index.html.

Fig. 2.10. Photos of 17th Street Canal breach and bridge destroyed over Bay St. Louis, Mississippi, are courtesy of the National Oceanographic and Atmospheric Administration from www.katrina.noaa.gov/index.html. Photo of car left behind in Mereaux, Louisiana, was taken by Barry Keim. Photo of flooding at childhood home of Barry Keim was taken by Arnold Crabtree and is reproduced by permission.

Fig. 2.11. Top photo of barge and bus was taken by Patrick Hesp and is reproduced by permission. Bottom photo was taken by Robert Muller.

Fig. 2.12. Photos of scenes from around the Ninth Ward of New Orleans were taken by Barry Keim and Robert Muller.

Fig. 2.13. Photos from St. Bernard Parish after the ravages of Hurricane Katrina were taken by Barry Keim.

3. COUNTING TROPICAL STORMS AND HURRICANES OVER THE GULF OF MEXICO

Fig. 3.2. Spatiotemporal pattern of tropical storm, hurricane, and severe hurricane strikes along the Gulf and Atlantic coastlines from 1901 through 2007 is adapted from Keim et al. (2007).

Fig. 3.4. Severe hurricane tracks of the 1950s and 1960s is redrafted from Blake et al. (2007) and was first published in Keim and Muller (2008).

4. HURRICANE BASICS

Fig. 4.4. Image of Hurricane Andrew in the Gulf of Mexico on 25 August 1992 is from the National Wetlands Research Center, found at http://www.nwrc.usgs.gov/images/andrew.jpg.

Fig. 4.5. Tropical storm and hurricane tracks of the North Atlantic Basin from 1886 through 2006 is adapted from Neumann (1993) and was first published in Keim and Muller (2008).

Fig. 4.6. Frequency of hurricanes and tropical storms in the North Atlantic Basin over the course of a hurricane season is modified from the National Hurricane Center version as found at http://www.nhc.noaa.gov/pastprofile.shtml.

Fig. 4.7. Chronology of the loop current is courtesy National Oceanographic and Atmospheric Administration.

Fig. 4.8. Track of Hurricane Katrina in relation to the location of the loop current is modified from and used courtesy of NASA Jet Propulsion Lab and the University of Colorado.

Fig. 4.9. Track of Hurricane Rita in relation to the location of the loop current is modified from and used courtesy of NASA Jet Propulsion Lab and the University of Colorado.

5. MEMORABLE GULF HURRICANES

Fig. 5.2. Public domain photo of the only house to survive the Cheniere Caminada Hurricane is found at 2theadvocate.com.

Fig. 5.4. Photo of the Poland Station Streetcar Barn in New Orleans is from the "New Orleans Streetcar Album" of H. George Friedman, Jr., University of Illinois at Urbana-Champaign.

Fig. 5.6. Photo of train washed off its tracks was provided by the State Archives of Florida.

Fig. 5.9. Photo of flooding in New Orleans resulting from Hurricane Betsy was taken by R. Vetter of the American Red Cross, provided courtesy NOAA/Department of Commerce Photo Library.

Fig. 5.12. Photos of the Richelieu Manor Apartments in Pass Christian, Mississippi, were provided courtesy NOAA/Department of Commerce Photo Library.

Fig. 5.18. Photo of tornado damage in LaPlace, Louisiana, associated with Hurricane Andrew was provided courtesy NOAA/Department of Commerce Photo Library.

6. HURRICANE HISTORIES

Fig. 6.5. Photo of beach scene from Pensacola, Florida, was taken by Ricky Hudson of CB4GO.com and is reproduced with permission.

7. ENVIRONMENTAL AND SOCIOECONOMIC IMPACTS

Fig. 7.1. Photo of Gulf Shores beach resorts was taken by Ricky Hudson of CB4GO.com and is reproduced with permission.

Fig. 7.4. Photo of gridlock during the evacuation before Hurricane Rita was taken by Ed Edahl and is from the FEMA Photo Library.

Fig. 7.5. Chart of Tropical Storm Allison rainfall is modified from the Tropical Storm Allison Recovery Project, http.//www.tsarp.org/images/map6.gif.

Fig. 7.7. Hurricane Camille rainfall pattern is adapted from NOAA Hydrometeorological Prediction Center's depiction at www.hpc.ncep.noaa.gov/tropical/rain/camille1969filledrainblk.gif.

Fig. 7.9. Hurricane Katrina peak storm surge as estimated by the ADCIRC Model is adapted from Executive Summary, U.S. Army Corps of Engineers Interagency Performance Evaluation Task Force (IPET) Report.

Fig. 7.10. Depiction of surge along Mississippi and Alabama coasts during Hurricane Camille is adapted from United States Corps of Engineers (1970).

Fig. 7.11. Chart of coastal land loss in Louisiana from 1870 projected to 2090 is used courtesy of the Louisiana Water Resources Research Institute, Louisiana State University.

Fig. 7.12. Data for locations of oil and gas platforms along the U.S. Gulf Coast are from National Weather Service, PowerMap, and the U.S. MMS.

Fig. 7.13. Photo of rig washed aground after Hurricane Katrina is reproduced courtesy NOAA/Department of Commerce Photo Library.

Fig. 7.14. Photos of damage to the oil and gas infrastructure are from the LSU Center for Energy Studies.

Fig. 7.15. Drawing of New Orleans in cross-section is from the U.S. Army Corps of Engineers.

8. THE FUTURE

Fig. 8.1. Chart of hurricane-observing platforms over time is modified from Neumann et al. (1993).

Fig. 8.2. Time series of the Atlantic Multi-decadal Oscillation and of the AMO detrended for global warming are adapted from Vermani and Weisburg (2006).

Fig. 8.3. Average global temperature time series as depicted by the National Climatic Data Center and comparison of severe hurricane tracks were provided courtesy Tropical Meteorology Project, Department of Atmospheric Science, Colorado State University.

Fig. 8.4. Comparison of severe hurricane tracks over two 13-year periods was provided courtesy Tropical Meteorology Project, Department of Atmospheric Science, Colorado State University.

EPILOGUE

Fig. E-1. Photos of damage to the home of Woody Keim were taken by Woody Keim and are reproduced with permission.

Fig. E-2. Photos of Hurricane Ike damage on the Bolivar Peninsula were taken by Robert Muller.

INDEX

Page numbers in italics refer to illustrations.